DATE DUE

DEMING

The Way We Knew Him

DEMING

The Way We Knew Him

Frank Voehl
Strategy Associates, Inc.
Coral Springs, Florida

St. Lucie Press
Boca Raton Boston New York Washington, D.C. London

Library of Congress Cataloging-in-Publication Data

Deming : the way we knew him / [edited by] Frank Voehl,
p. cm.
Includes bibliographical references and index.
ISBN 1-884015-54-9 (hardcover)
1. Deming, W. Edwards (William Edwards), 1900– –Contributions in management. 2. Management. 3. Quality control. I. Voehl, Frank, 1946– . II. Deming, W. Edwards (William Edwards, 1900–
HD31.D4223 1995
658—dc20 94-46387
CIP

International Standard Book Number 1-884015-54-9
Library of Congress Card Number 94-46387
Printed in the United States of America 2 3 4 5 6 7 8 9 0
Printed on acid-free paper

DEDICATION

To my parents,
Frank and Ruth Voehl,
who have always been quality examples
of quality people leading quality lives in action.
Thank you for teaching me how to live.

To my daughter,
Danna,
who was a quality example
of what could be accomplished and handicaps overcome
in a shortened lifetime of just ten years.
Thank you for teaching me how to love

To my mentor in quality at FPL,
Joe Collier,
who was a true unsung pioneer
in the U.S. quality movement
and who died as did Dr. Deming,
gravely ill yet actively pursuing the Quality Grail.
Thank you for teaching me how to lead.

CONTENTS

FOREWORD

Dr. Deming would often say that he was very fortunate to only study under great men. Any of us who had the chance to attend a Deming seminar, or read his work, can be justified in saying the same. Anyone anxious to learn found Dr. Deming a willing teacher. I remember how his eyes lit up with excitement when I asked him where I could get books by Shewhart to study. Dr. Deming said the following about his friend and mentor, Walter Shewhart, at the time of Shewhart's death in 1967:

> ...he was always glad to help anyone. Actually, he never thought of himself as helping anyone; he was simply glad to talk and absorb thoughts from anyone who was genuinely struggling to improve his understanding of the statistical method—interchanging ideas was his way to put it.

These two men must have been cut out of the same cloth. Those of us who have been Dr. Deming's diligent students have found a new zest for learning and improvement. Just like all great explorers, he was not afraid of the unknown and viewed the uncertainty of the unknown and the future as an opportunity to create new knowledge. I will think of him as I learn new and wonderful things in the years to come.

John Peterson
1993 Deming Forum President

Among Dr. Deming's several great contributions, I think that he played a leading role in applying mathematical statistics to quality control for the practicality of its usage, as well as transferring its practical methods to quality control specialists and managers. His other great contribution was to make the Japanese people gain self-confidence and lead us to a bright future. He made us believe that there would be a possibility to improve quality even amidst the disaster after the second world war.

In the United States, Dr. Deming's thoughts and philosophy, including his Fourteen Points, affected executives beyond measure—inscrutably. I regret that you have lost your greatest leader. We shall always remember him as one of the leading quality control specialists in the world.

Dr. Noriaki Kano
Department Head and Professor
Department of Management Science
Science University of Tokyo
(from a letter sent by Dr. Kano
and provided by Lou Schultz)

PREFACE

The day I heard that Dr. Deming had died, I was on my way from Miami International Airport with my friend Hana Tomasek. She was filling me in on a trip that she had just made to the Czech Republic and told me of her never-to-be-realized dream of bringing Dr. Deming to her native country. As we talked, the idea for a book of remembrances about the life and person of Dr. Deming filled my thoughts. I called Myron Tribus about the idea, and he was quick with his encouragement and personal support, as was Mary Walton. Howard Gitlow provided many valuable contacts, and before long I had a checklist of some twenty names who would provide the basis of the text.

With this early encouragement, I began to formulate the purpose: to describe the impact and person of Dr. Deming in terms of certain significant attributes, characteristics, and interpretations of titles that have been spoken of him and to present these contributions in a comprehensive manner without being overly repetitive. The general aim of the book is to learn as much as possible about Dr. Deming and his work from the names and titles given to him. His deeds, character, and position in life are summed up by some of those around him—those who knew him in a special way, those who felt his impact and influence, and those whose voice to be heard will bring a special meaning and insight into the man and the legend.

With this purpose in mind, I began the journey of building this book along with the other authors that give life to these pages. The problem then became how to organize and integrate all this material without it being repetitious or unfocused. After taking two or three blind alleys, my wife, Micki, helped me think it through as usual, and we came up with the idea of organizing the material around the Fourteen Points. This turned out to be a fortuitous move for many reasons. First, it added to the consistency of the material, and it allowed Deming to interpret Deming in an unusual way. It also provided for a richness and logical flow that could not be present in any of the original contributions.

Beginning in Chapter One, Mary Walton provides real insight into just what adopting the new philosophy means. Her journey took her through the Fourteen Points with Dr. Deming on a line-by-line and point-by-point basis. In Chapter Two, we learn from Dave and Carole Schwinn the real issues involved in leadership and where Dr. Deming was a pioneer and a genius at the same time. In Chapter Three, Myron Tribus gives us a bird's-eye view of what the systems approach looked like in the outcomes of Dr. Deming's life.

In Chapter Four, we learn about training from an interesting source, a husband and wife team who first encountered the point about instituting training in a faraway land twenty years ago. This theme is continued in Chapter Five, where Nancy Mann gives us wonderful insights into Dr. Deming as the holder of the keys to excellence, where modern methods of education and self-improvement are vital. Chapter Six completes the cornerstones of the Deming philosophy with a look at constancy of purpose through the statistician's eyes, as provided by an old friend of Dr. Deming and one of his first graduate students, Ernie Kurnow.

In Chapter Seven, Homer Sarasohn looks closely at the ineffective practice of price tag awards and gives us insight into the early days of the Japanese transformation using SPC and the principles of Scientific Management, before Dr. Deming appeared on the scene. In Chapter Eight, Howard Gitlow discusses the *michi* of the statistician, our cosmic footprints in the sands of life, in a unique way that makes the concept come alive.

In Chapters Nine through Fourteen, the contributors present their unique insights into the Fourteen Points. Breaking down the barriers, especially at the community level, is deftly described by Mary Ann Gould and Maureen Glassman. In Chapter Eleven, Lisa McNary, the last of a long line of graduate students, describes her feelings for Dr. Deming in terms of lighting the candle to drive out fear. Gerry Glasser's tribute takes us back twenty years to Dr. Deming's Greenwich Village Apartment. In Chapter Twelve, I discuss the curse of the prize, which befalls organizations that chase after glory. In Chapter

Thirteen, Lou Schultz brings home the concept of pride of workmanship in a new way. In Chapter Fourteen, Father Peters presides over a requiem for the heavyweight of the quality movement and shows how the transformation is everyone's business in the universal brotherhood of all mankind.

The book closes with a look at the Seven Deadly Diseases, with which Dr. Deming was at constant war. Some might ask why Profound Knowledge is not covered, as it represents a substantial portion of the Deming philosophy, especially in the last five years or so. My reply is that we first must gain a better understanding of the basics of the Fourteen Points, which is one of the objectives of this book. To that end, I hope that we have succeeded, even in some small way.

The promise of this work is that you will enjoy reading it half as much as I enjoyed pulling it all together. I learned a lot and grew a lot. The only thing left to say is, "Thank you, Dr. Deming, one more time." And thank you, authors, for your moving testimonies to the master of the quality movement, W. Edwards Deming: pioneer, genius, statistician, guru, mentor, counsellor, legend, friend, herald angel, man among men!

Frank Voehl

CREDITS

We are indebted to the following sources for the use of their material: *Out of the Crisis* by W. Edwards Deming (Center for Advanced Engineering Study, Massachusetts Institute of Technology, Cambridge, 1989), *The Keys to Excellence: The Story of the Deming Philosophy* by Nancy R. Mann (Prestwick Books, Los Angeles, 1989), *The Deming Management Method* by Mary Walton (Dodd, Mead & Company, New York, 1986), *Deming Management at Work* by Mary Walton (G.P. Putnam's Sons, New York, 1990), *The Stigma of Genius* by Kincheloe, Steinberg, and Trippins (Hollowbrook Publishing, Durango, Colorado, 1992), "By His Works Shall Ye Know Him" by Myron Tribus (from a 1993 speech to the Society of Professional Engineers, Pittsburgh), "The *Michi* of the Statistician" by Howard Gitlow (University of Miami, 1994), The Fourteen Points of W. Edwards Deming by Frank Voehl (Florida Atlantic University, 1993; Strategy Associates, 1994), "The Deming Prize" by Frank Voehl (*South Carolina Business,* 1991, special issue entitled "In Pursuit of Total Quality," published by the South Carolina Chamber of Commerce), "The Sarasohn Chronicles" by Homer Sarasohn (Scottsdale, Arizona).

PROLOGUE

On June 24, 1980, NBC aired the now-famous documentary entitled "If Japan Can...Why Can't We?" Its purpose was to discover how the Japanese had risen from the ashes of World War II to become a self-sufficient leader in quality in less than fifteen years and a world economic power in thirty years. By 1980, the United States was importing about $30 billion in Japanese goods, which consisted of manufactured items such as automobiles and electronics products. On the other hand, the United States was exporting about $20 billion worth of goods to Japan, mostly in the form of raw materials such as seed, soybeans, lumber, coal, non-ferrous metals, and so forth. At the time, the Japanese were accused of not playing by the rules and of dumping their products, copied from U.S. technology, onto American markets.

The NBC documentary changed some of these perceptions by probing the facts behind the story. It featured an eighty-year-old statistician named W. Edwards Deming, who had taught the Japanese the art and science of Statistical Process Control. These were the same methods that had enabled Western management—the Allies—to manufacture superior weapons and bombs which were eventually used to win the war. As Mary Walton reported in *Deming Management at Work,* "the NBC White Paper documentary, more than any single event, set America on a new course towards quality, with Dr. Deming at the helm. Among the companies that asked for his help were Ford, General Motors, and Nashua Corp. among others. Some would fall by the wayside as the rigors of the task would become evident and others take their place. Attendance at the Deming four-day seminar would swell from a dozen or two during the 1970s to hundreds. And during the course of all this, he was merciless in his condemnation of Western management."

During the 1980s, the American quality movement took on a life of its own. Teams at all levels became the norm, at least in successful organizations, working on quality issues consistent with corporate goals. At the start, quality was the mission of a "shadow" organization—the Quality Department—with its army of facilitators and trainers who were ready to fix the system. As the movement matured, continuous improvement became the way of life. Dr. Deming's Fourteen Points became the hub of the quality movement, and his Theory of Profound Knowledge became the spokes of the wheel. None of this was difficult to understand—nor easy to make happen.

Just who was this Dr. Deming, the eighty-year-old statistician who was "discovered" in 1980 and who helped to revolutionize the quality movement in Japan as well as the Western world? How did he come to be recognized late in life as "the father of the third wave of the industrial revolution"? The answers are contained in the pages of this book, which is constructed as a series of chapters that relate to Dr. Deming's Fourteen Points. Each chapter begins with a brief explanation of the Deming philosophy, followed by a tribute written by a well-known expert or other individual close to Dr. Deming. By building the book around the Deming philosophy, a deeper understanding of his genius begins to emerge.

As I struggled with the organization and consistency of this presentation, the value of centering the work around the Fourteen Points became apparent. Although not necessarily the latest thinking among his disciples and inner circle, the Fourteen Points remain a unique contribution and a testament to his true genius and his ability to weave disparate elements and concepts into a coherent system of knowledge. While his Theory of Profound Knowledge may be considered more advanced by some, his Fourteen Points have a more universal appeal. Add to this the Seven Deadly Diseases, and the result is a rich collection of testimonies that give special insight into the life, times, and unique genius of W. Edwards Deming.

As pointed out by Kincheloe, Steinberg, and Tippins in *The Stigma of Genius,* the innovative concept of "genius (or what is generally perceived as genius) regardless of time or place, evokes negative reactions. Innovators are flies in the ointment of the dominant culture. They are members of the monkey wrench gang who disturb the drowsy status quo. They reject the comfort of consensus, as they point and giggle at the emperor's nakedness." In other words, they teach us to redefine genius and intelligence.

Einstein was one of Dr. Deming's role models for disturbing the status quo. Dr. Deming told the story of when Einstein returned from a trip abroad after winning the Nobel Prize in physics. As the scientist debarked from his ship in New York harbor, he was surrounded by a crowd of reporters. One of them asked him a question about the speed of sound, to which he replied, "I don't know." The reporter was amazed and blurted out, "But you're the smartest man in the world. Why don't you know?" To which Einstein replied, "I don't need to know. I can look it up any time that I need the information."

The journalist's question reflects the basic problem with the twentieth century view of intelligence, which is measured by one's capacity to store information. Einstein possessed a form of intelligence that moved beyond normal ways of formal thinking, and in doing so he often disturbed the drowsy status quo. Dr. Deming was also one of those disturbers, one of the leaders of the "monkey wrench gang." And like Einstein, he was one of the gifted few who had the courage to tell the emperor the truth about his new clothes.

Let's begin our journey into the mind and heart of one of the greatest Americans of this century—W. Edwards Deming, a man for all seasons.

Part One

THE CORNERSTONES

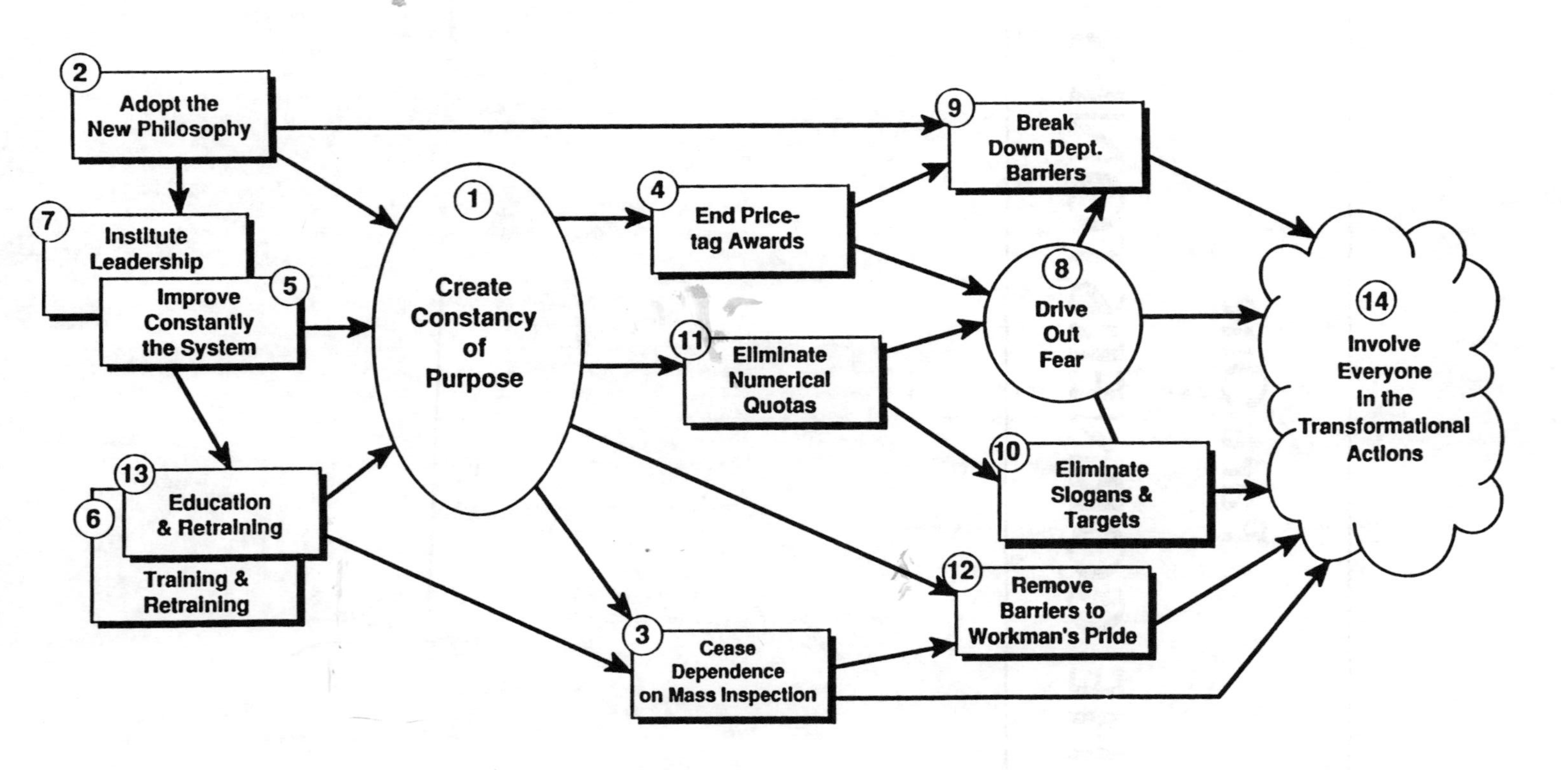

Deming's Fourteen Points System Diagram

Chapter One

ADOPT THE NEW PHILOSOPHY

"America must adopt the new philosophy.
We are in a new economic age.
Western management must awaken
to the challenge, must learn to take
responsibility and leadership for change."

Dr. Deming often reminded us that quality is like a new religion. As such, it requires a complete transformation of mind and heart—a *metanoia,* as he sometimes called it—starting with management. He constantly reminded us that between thirty and forty percent of an American business may be lost in what Dr. Armand Feigenbaum refers to as the so-called "hidden plant," that is, a factory that exists to perform rework on unsatisfactory parts, reinspect or retest rejected parts, and rework field returns. In other words, the part of the business that operates to process the defects, rework, and scrap—the junk.

The problem begins with attitude—on the part of both management and the company. Accepting defective materials, poor workmanship, and inattentive service eventually leads to poor quality and lower productivity, and only a dismantling of the structures of current management practice will do, Dr. Deming taught. He often said that we have to undergo a complete demolition of the American style of management which had spread in some form or another to the entire

Western world. Best efforts and hard work will not suffice, nor will new machinery, computers, automation, or gadgets, he argued. The central idea that he taught and believed is that we are being ruined by best efforts put forth with the best of intentions but without the guidance of a new philosophy—a theory of management—for the optimization of a system. One of his favorite expressions was, "there is no substitute for knowledge."

In his seminars and lectures, he pointed out over and over again that the cost of living is inversely dependent upon the goods and services that a given amount of money will buy. On more than one occasion, I had the opportunity to discuss this paradox with Dr. Myron Tribus, often at 1 p.m. in his study, with a final pot of coffee on the stove. Myron helped to explain Dr. Deming's ideas on this subject as follows: In a competitive situation, profits for many companies come by mortgaging the future and cutting the costs of R&D, training, and other forms of competitive marketing research. In many cases, management is convinced that the installation of some new and complex equipment is the solution to their productivity problems. Paradoxically, however, by increasing the capital investment per worker, productivity is often lowered, instead of raised, because of poor management. This is due to the tremendous influence and leveraging effect of the management cycle. The greater the capital investment, the lower the productivity—a vicious cycle.

To help us better understand this principle, in Chapter 6 of her first book, *The Deming Management Method,* Mary Walton relates a story that Dr. Deming tells about a "beer manufacturer who boasted that he had no problem with beer cans because his suppliers replaced those that were defective. It never occurred to him that he was paying for the defective cans, because their cost was included in the wholesale price. Deming strongly believed, and lived his belief, that consumers of goods and services end up paying for delays and mistakes, which ultimately reduce their standard of living and that of society as a whole."

Mary Walton first heard of Dr. Deming on a trip to Japan in the early 1980s while researching a story for the *Philadelphia Inquirer.* The subject was workers at Kawasaki Heavy Industries, which had recently won a contract to build subway cars for the Philadelphia mass transit system. She tells how she thought that Deming had passed away years ago or drifted out of sight after educating the Japanese, as his name was not exactly a household word in America at the time. She was surprised to find him alive and well and coming to Philadelphia in 1984 to conduct a four-day seminar for Maureen Glassman of PACE. The story of that first meeting and the relationship that ensued is a compelling one, as only Mary can tell it.

A Journalist Remembers Dr. Deming

by Mary Walton

No, Cecelia Kilian told me, she could not possibly arrange an interview with her boss, W. Edwards Deming. He simply did not have time. No, I could not ride the plane with him from Philadelphia to San Diego. He needed to rest and reflect. Yes, however, I could sit through his seminar, although I would have to arrange it with George Washington University and pay whatever they asked.

Didn't these people know how the game was played? I was a reporter for the Sunday magazine of *The Philadelphia Inquirer,* whose story would reach a million households. The magazine was under constant pressure from press agents seeking coverage such as I proposed, that is to say, a long profile, possibly even a cover story. There were people who would do anything for this kind of publicity. And I couldn't even get an interview?

This was my introduction to the unorthodox world of Dr. Deming and the beginning of a ten-year association until his death in December 1993. Knowing him would change my life, both personally and professionally, as it did that of so many others. I would write two books with his support and become known, to my amazement, as an expert in quality.

The beginning seemed anything but auspicious. It was January 1984, and Dr. Deming had emerged just over three years earlier from the obscurity that had dogged him for decades in the United States despite his distinguished reputation in Japan. He was coming to Philadelphia at the invitation of the Chamber of Commerce to give

---

Mary Walton is a well-known journalist and author who has published two books about the Deming Method. She holds an MBA from Harvard University and is considered by many to be a quality guru in her own right.

a three-hour preview of his four-day seminar scheduled a few months hence in the spring of 1984. Deming arrived in Philadelphia with a Ford Company vice-president and several other executives to bear witness to the advances his method offered. After Philadelphia, his schedule called for a seminar in San Diego, which I had made arrangements to attend. My mission was to write a story in advance of his spring appearance, which would kick off the quality and productivity campaign that later became institutionalized as PACE (the Philadelphia Area Council for Excellence).

When Dr. Deming arrived in Philadelphia, I introduced myself and said I would see him in San Diego. "That's wonderful," he boomed. "Why don't you ride the plane with me?" So there, Ceil!

I would come to understand that in turning down my request for an interview, his loyal secretary Ceil was merely trying to protect him from a stream of desperate petitioners who waylaid him at every turn. for every reason imaginable for advice on their business, not to mention others who sought his endorsement for any number of projects.

Airborne for California, I managed to sit at Deming's side in the first-class section for two hours, until the flight attendant directed me back to my coach seat. He introduced me to the principles of Statistical Quality Control, expounding on common causes and special causes of variation, management responsibility for quality, the futility of performance ratings, and so on. I had no idea what he was talking about.

In San Diego things became somewhat more clear. I wrote the story, a cover story as it turned out, "W. Edwards Deming Wants to Make America Work Again." It was published on March 11, 1984 and generated some fifty calls and letters the first week, which was more reader response than anything I had written to date. (The previous record-holder was an article entitled "Why Women Hate Their Gynecologists." In both cases, it was clear I had touched a chord.)

Thus encouraged, I proposed to Dr. Deming that we do a book on his method. "By W. Edwards Deming and Mary Walton" was how I saw it. He had liked the article and thought a book was a good idea. I contacted an agent, who said that the proposal was worth a $200,000 advance. To a $40,000-a-year reporter, this was a fortune. I called Dr. Deming with the good news. "An agent? I never deal with an agent," he rumbled in that deep, raspy voice I was getting to know so well. I learned that he had his own publisher, thank you very much, an academic press, and that in our discussion he had meant only that he would help me. Moreover, he was then in the midst of his own book, the forthcoming *Out of the Crisis*.

For a book not by but about Deming and his method, the price dropped to $15,000. Even at that, many publishers turned down the proposal. "Too smokestacky," one said, voicing the common objection. This was the mid-1980s, and quality was something that seemed relevant only to cars, electronics, steel, and other industrial products.

Dr. Deming made good on his pledge of support. I attended additional seminars and traveled with him to several clients, including Ford. He loaned me a mimeographed copy of his diaries from Japan and provided introductions to people he thought would be helpful. He also granted his scarcest resource: time.

My visits to him on occasional Saturdays always followed the same pattern. I would arrive at his Washington home by train from Philadelphia in late morning. We would spend several hours talking, and then we would go to the Cosmos Club for an early dinner. The Cosmos Club was a genteel men's establishment with an aging clientele and a Southern ambiance from the days when a more insular Washington mirrored its geographical location. Black waitresses in white aprons served popovers before dinner. Dr. Deming pressed martinis on his guest and, after dinner, hazelnut ice cream. Once we were accompanied by his wife, Lola and my daughter, Sarah, a budding teenager. Afterward, he often asked about her by name. I marveled that he remembered.

Despite his failing eyesight, in those days he still piloted his massive 1969 white Lincoln Continental through Washington's snarled streets with blissful disregard for other traffic. After one harrowing ride in the rain, I called his daughter Linda Haupt. "You've got to stop him from driving," I told her. "I've tried," she wailed. It was not easy even for a family member to make Dr. Deming do anything. One time we rode the bus. "I ride for twenty-five cents," he said with satisfaction.

Following dinner, I would take a cab to Union Station and a train back to Philadelphia. Dr. Deming always called the next day to make sure I had arrived home safely.

In writing the book, the first hurdle was to secure Dr. Deming's approval for my interpretation of his philosophy. We started with what it means to "Adopt the New Philosophy." We proceeded word by word, chapter by chapter, through each of his Fourteen Points, one by one, and then on to the Deadly Diseases and Obstacles. The problem was that he was no more patient with me than with those unfortunate people who raised their hands at his seminars, only to be unceremoniously dismissed. "Weren't you listening?" he would sometimes say, or, "I don't understand what you mean." Along the way, I discovered that Dr. Deming disliked certain words intensely, one of which was "implement." "A farming tool," he scrawled in the margin where he had crossed it out. He disapproved of contractions, and eliminated them throughout the manuscript—in his and other people's quotations.

Once we had an acceptable manuscript, I faced another hurdle. His publisher, Massachusetts Institute of Technology's Center for Advanced Engineering Study, then under the direction of Myron Tribus, would not grant permission to quote from his book of seminars, claiming copyright of both. Perhaps they believed mine to be a competitive work. Not until my publisher, Dodd, Mead & Company, halted publication of my book, plunging me into despair, did Dr. Deming step in and personally sign the permission forms. Then a difficulty arose with Dr. Deming's foreword, which was written in old-fashioned prose. The editor drastically revised it against my advice. An irate Dr. Deming withdrew the foreword. With it went his stamp of approval for the book. The editor immediately yielded, but it was I who had to beg for its reinstatement.

As the book neared publication, I was beset by other difficulties, one of which was that Dodd, Mead was in financial trouble. The company first fired my editor and then the art director and marketing director. Within a year of the book's appearance in 1986, the whole company went belly up. Fortunately, Putnam Publishing Group bought *The Deming Management Method.* Once rescued, the book escalated in sales, reflecting its subject's growing popularity.

Three years later I approached Dr. Deming for help with a second book on quality, a study of companies that were laboring to follow his teachings. "People would read my first book, then go to your seminars, but they feel lost," I told him. "They say, 'This sounds wonderful, but tell me where it's working. Who else can I talk to? Where can I find out what companies are adopting the new philosophy?'"

Dr. Deming snapped, "Well, that's a joke. There aren't any." He conceded, however, that "bits and pieces" of his philosophy were finally being implemented. We started with Florida Power and Light, Hospital Corporation of America, Bridgestone Tennessee (Tri-Cities,

Tennessee), the U.S. Navy, and Globe Metallurgical Inc. The book became known as *Deming Management at Work*. As he did for our first book, he did for this one also. He ended the foreword with a challenge:

> The change required is transformation, change of state, metamorphosis, in industry, education, and government. The transformation will restore the individual by the abolishment of grades in school on up through the university; by abolishment of the annual appraisal of people on the job, MBO, quotas for production, incentive pay, competition between people, competition between divisions, and other forms of sub-optimization. The transformation is not stamping out fires, solving problems, nor cosmetic improvements. The transformation must be led by top management.

We were off and running. As I listen to his voice on that tape from 1989, I feel even more sadness than I did on learning of his death. I had not seen much of him in recent years. Our last encounter was in July 1993, five months before he died, at a seminar in Detroit. I arrived unannounced and I don't believe he recognized me. But thinking of him again brings back memories of his brilliance, his eccentricity, and his kindness to a journalist who had stumbled into the new philosophy of the quality movement and whose life would never be the same.

> ***A** conjurer may pull a rabbit out of a hat, but he cannot pull quality out of a hat. The biggest problem that most any company in the western world faces is not its competitors nor the Japanese. The biggest problems are self-inflicted, created right at home by management that are off course in the competitive world of today. To get back on track a new philosophy is needed—a philosophy of continuous improvement, starting with management.*
>
> W. Edwards Deming
> From the Foreword to
> *The Deming Management Method*

Chapter Two

INSTITUTE LEADERSHIP

---

*"America must institute leadership.
The aim of leadership should be to help people
and machines and gadgets do a better job.
Leadership of management is in need of overhaul,
as well as leadership of production workers."*

Dr. Deming taught that the job of the supervisor is not to tell people what to do or to punish, but to lead. To him, leading consisted of helping people do a better job and learning by objective methods who may be in need of individual assistance. He also believed that it is management's responsibility and role to discover and remove the barriers that prevent workers from taking pride in what they do.

He believed that most supervisors hinder workers in doing a job properly because, in many cases, they are hired right out of college and lack experience in that they have never done the job. In days of old, American foremen knew the work and knew how to identify the employees who were in need of help. Their goal was to find them and help them. All that has changed in America today. The changes began after World War II, when the United States got away from the things that made it great and won the war—superior weapons and bombs, strong military leadership, and loyal citizens.

Management has failed in the United States, Dr. Deming repeatedly stated at his seminars. The emphasis here is on the quick buck, while the emphasis in Japan is on planning for the decade ahead. The next quarterly dividend is not nearly as important as the existence of the enterprise five, ten, or twenty years from now. A key requirement of innovation is the belief that there will be a future.

According to the Deming philosophy, the job of the foreman as a leader is to provide direction to the production workers and to report any problems with which the workers may have to deal. Accomplishing this leadership role, Dr. Deming taught, may require the use of control charts and statistics. Rather than acting as judges, foremen need to behave as facilitators, helping workers to build in quality and increase productivity.

After the war, starting in 1945, the Japanese learned these lessons well, and their leadership practices have become a model. Dr. Deming saw firsthand what happens when management is not involved from the very beginning, and when he had the chance to go to Japan in 1949, he was determined not to make the same mistakes that resulted in a setback for him in America after the end of World War II.

Dave and Carole Schwinn have known Dr. Deming since the early 1980s and have heard his firsthand account of the Japanese transformation, starting with leadership. As expanded upon by Homer Sarasohn in Chapter 7, Dr. Deming saw the opportunity and seized the moment. His leadership practices became the model to follow. The Japanese did and the Americans did not—and the rest is history.

W. Edwards Deming: Personal Remembrances of Leadership

by Carole and Dave Schwinn

The Father of the Third Wave of the Industrial Revolution—That is what Bill Conway, then chairman of the Nashua Corporation, called Dr. Deming in the historic 1980 NBC special "If Japan Can...Why Can't We?" *U.S. News and World Report* called his contribution one of the nine most significant hidden turning points since the discovery of America. It will take years to fully understand Dr. Deming's role in history, but to many of us he was a model, teacher, and friend—the essence of a leader.

Dr. Deming himself described what it meant to be a model. In a seminar at Ford Motor Company, he said, "Did you ever try to write instructions on how to walk? Very difficult; yet people learn to walk by watching others." By watching him, one could learn lessons for a lifetime of leadership.

First, he modeled the leadership attribute of *generosity*. It is well known that he declined payment for much of the early teaching he provided to the Japanese. That money became the funding for Japan's Deming Prize. The Deming Prize process itself, of course, became Japan's model for improvement and economic transformation. Later, his generosity extended to those of us who were so eagerly trying to learn and apply his teachings. Dr. Myron Tribus relates that Dr. Deming never failed to respond when Tribus asked for a small donation to help someone trying to take the theories to practice, particularly in education.

An early experience of our own with Dr. Deming illustrates that he was also a model of *sensitivity and attention to the needs of his followers*. The first time Carole (then Hannan) met him was in the summer of 1983 at one of his famous four-day seminars at Ford Motor Company. David had by that time worked with him at Ford for almost

---

Professors Carole and Dave Schwinn are involved in the quality movement through Jackson Community College, Jackson, Michigan, as well as the Transformation of American Industry (TAI) project.

two years. In anticipation of meeting Dr. Deming, Carole wrote him a letter describing the work that was going on at Jackson Community College to try to educate local auto suppliers in his principles and practices. At the first break, on the first day of the seminar, David escorted Carole to meet Dr. Deming and said, "Dr. Deming, this is Carole Hannan." To Carole's great chagrin, he promptly turned and walked away! We watched as he walked to his table and fumbled with some papers. Then he turned and came back to us with a letter in hand. "Carole," he said, "I received your letter and responded. But I feared that you would not receive it before the seminar, so I brought along a copy to hand to you." Needless to say, we never again questioned his sensitivity and attention to our needs. The letter he handed to Carole that day illustrates Dr. Deming as a model of a leader who *challenges assumptions.*

Another letter, received only two weeks later, serves as just one example of Dr. Deming as a *coach.* In the letter, he suggested to us how we could further improve our teaching in Jackson. "The important thing about the Jackson Quality Group," he wrote, "is to study the first four chapters of my notes entitled 'Quality, Productivity, and Competitive Position.' Unless top management in the community understands the 14 points and the deadly diseases...there would be little hope of any accomplishment worth noting." When we sent further documentation of our work in Jackson several months later, he continued his coaching role in the ensuing months.

Dr. Deming's leadership extended to his modeling the importance of *celebration and play.* He suggested that we visit Japan with Dr. Tribus and a group from the Philadelphia Area Council for Excellence (PACE) on the occasion of the thirty-fifth anniversary of the Deming Prize (see next letter). When we arrived in Tokyo, we joined Dr. Deming and others in a delightful dinner at Chingano Gardens. Before the meal was served, he suggested to the small group at his table that we each take the paper drink coaster in front of us, sign it, and circulate it for everyone else's signature. Thus, he said, we would all have some tangible remembrance of celebrating together our first night in Japan. That coaster still graces an honored spot in our home and in our hearts.

In later years, we began to gather greetings to Dr. Deming from those who were impacted by his work. We would then send them to him in a package on his birthday. One such greeting was from Dr. Myron Tribus, who early on recognized Dr. Deming's leadership model for *experimentation or lighting fires* wherever he could, all the while recognizing that not all of them would burn brightly.

Dr. Deming was also the model of *appreciation*. In response to the greeting cards on his birthday, he wrote:

> I thank you for birthday greetings and for your wonderful letter. It is a treasure. You do wonderful work. I shall read your letter again and again—already three times. I send best greetings, remaining ever...
>
> Sincerely yours,
> W. Edwards Deming

Perhaps most significantly for many, he was a model of the leader as a *life-long learner*. He constantly questioned others about their work and engaging them in conversations about his own work and writings. He would make notes on scraps of paper and tuck them away in his calendar. Often the notes resurfaced as references in his books and articles. The Deming Library tapes were also an occasion for him to continue to learn from others, including Alfie Kohn, Russell Ackoff, and Michael Maccoby, among others.

Like all great leaders, Dr. Deming never failed to *give recognition to those from whom he learned*. His writings and his seminars were peppered with references to things he had read or heard from others. That recognition could be personal and immediate. While attending another seminar at Ford just six months before he died, Carole handed him an article she had just written about communities as "whole systems." Given his illness, she did not really expect him to read it, let alone respond. Just days after the seminar, however, the written response arrived, in which he said:

> I read the article you handed to me. It is very helpful to me. I wish for you all things good, and remain...
>
> Sincerely yours,
> W. Edwards Deming

Teaching by story and example, another leadership attribute modeled by Dr. Deming, often took a humorous twist. He loved to say, "All my examples are stupid. People learn from stupid examples." At one early Ford seminar, he was trying to make his point about "no true value." His voice boomed at the audience, "Did you know there is no true value for the speed of light? Galileo concluded that if the speed of light is not infinite, it is awfully damned fast."

Another time, he told the story of standing in a field one night as a youth and viewing Halley's comet with his father. He concluded the story by saying that his father told him that the comet would come back in eighty years. He paused dramatically and said, "It did." Assuredly, he was the only person in the room who had seen Halley's comet twice! Many, however, would remember the story and his point.

Dr. Deming's teaching style, at least in private, was often by Socratic method, or by *questioning*. An unforgettable personal example illustrates the profound impact of his methods. He examined early versions of training materials we were using and asked, "Why are you teaching histograms?" What he forced us to figure out—that histograms have no meaning in the absence of a stable system—seems perfectly obvious now, but it was not at the time. His challenge forced us to search out the answers for ourselves and to be sure that we found the answer to the real question he was asking. It was obvious that Dr. Deming was also *persistent* in making sure that his students learned.

At Ford, he always refused to help us with any "real" problems, preferring to *focus on theory and principles*. He referred us to consultant David Chambers for our "statistical" lessons. We had begun to question whether or not Dr. Deming had any "practical" skills at all, when a serious problem finally caught his attention. Ford's operations research people had been working for nearly a year on a rail transportation problem between one of the stamping plants and one of the assembly plants. They were about to give up when Dr. Deming heard about the problem. After listening to a description of the circumstances, he helped them to figure it out in ten minutes—with a pencil and a control chart.

Lest any reader imagine that we are unaware of Dr. Deming's reputation as a model of leadership as intimidation, cantankerousness, and abrasiveness, the following story about the thirty-fifth anniversary of the Deming Prize celebration may clear the air. His address on that occasion was severely critical, not only of American management, but also of Japanese management, and even the Japanese Union of Scientists and Engineers (JUSE). His message was a warning that he was seeing signs of infection from "American diseases" in Japanese business leaders. More and more joint ventures were occurring in which the practices of multiple source purchasing and performance appraisal systems were being incorporated. He wanted to make sure that the Japanese system of performance appraisal and the philosophy of working cooperatively with suppliers were not compromised. He was not gentle. Not one of the several hundred top Japanese executives walked away misunderstanding his message. The reactions of the executives at a later reception were also consistent. "He is our mother," they said, "he cares about our well-being and sometimes he must scold us."

Dr. Deming could be biting and critical, but probably by design. It seemed as though he needed to use whatever method he could to *communicate the urgency of his message.* He believed that we are in crisis and that there is no time to make the critical transformation required. He was impatient and he was, after all, only human, as are all leaders. For those of us who were trying to learn, it was necessary to look beyond the not always consistent style to the message.

Another personal story may further clarify his public style, which was often misinterpreted as inconsistent and rude. By 1986, we became involved with Myron Tribus in helping to launch and support community-based quality initiatives. David was asked to give a presentation on our early experiences at the first international Deming User's Conference, sponsored by the Ohio Quality Productivity Forum (OQPF). While David was preparing his slides, Dr. Deming walked in and took a front-row seat, which immediately threw David into a state of anxiety. The anxiety was justified when, perhaps five minutes into the presentation, Dr. Deming rose from his seat and said, "David, that is very interesting, but unimportant." David, of course, knew exactly what Dr. Deming was saying. David was talking about teams and tools and techniques, while Dr. Deming demanded focus on the transformation of American styles of management. David spent the rest of his allotted time attempting to help others see Dr. Deming's point, but not everyone understood that Dr. Deming was trying, not always gracefully, to be a *facilitator of profound learning.*

To us, Dr. Deming was, most importantly, a model of a *kind and loving friend.* Some would still say that those are not and cannot be attributes of a leader in today's business and industry. Others, however, are beginning to see that the new leader's role is all about relationships. Our own experience would support the former rather than the latter view.

We are not alone in having been blessed to be students of Dr. Deming. Our personal and professional lives, like those of many others, have been guided, shaped, and nurtured by his teaching and by his example.

The aim of leadership should be to improve the performance of man and machine, to improve quality, to increase output, and simultaneously to bring pride of workmanship to people; not merely to find and record failures of men, but to remove the causes of failure; to help people do a better job with less effort.

W. Edwards Deming
Out of the Crisis

Chapter Three

IMPROVE CONSTANTLY THE SYSTEM

"We must improve constantly and forever the system of production of goods and services, in order to improve quality and productivity, and thus constantly decrease costs."

All work is performed as part of a system. The discipline of the systems view of work was a recurring theme that permeated every part of Dr. Deming's teachings. He believed that quality must be built in, starting at the design stage. He taught that every product should be considered to be one of a kind because there is only one chance for optimum success.

Dr. Deming firmly believed that quality starts with intent, driven and fixed by management. This intent must then be translated into plans and specifications to produce products that are inspected and tested. Thus evolves the framework of the system of continuous improvement—"a cycle of never-ending improvement," as he called it.

One of my earliest conversations with Dr. Deming centered on this systems view of production. He believed that if we want to achieve a lasting culture in which continuous improvement and customer focus are a natural pattern, we must redesign the system to be consistent with the vision and values of the organization. Unfortunately, he said, the production system is always in a state of flux due to pressure from influences in the external, political, and technological environments. The situation is compounded because most organizations do not think through the impact of change within the system in any organized manner. In Dr. Deming's world, change occurs when the pain of remaining the same dysfunctional unit becomes too great and a remedy for relief is sought.

Just what is a system of continuous improvement, and how do we begin to define and understand its content? From the "exact science" days of the 1800s until the 1920s, the simple picture emerged in which specification, production, and inspection were considered to be independent of each other when viewed in a straight-line manner. Dr. Walter Shewhart taught Dr. Deming that the picture is entirely different in the absence of an exact science. When the production process is viewed from the standpoint that the control of quality is a matter of probability, then specification, production, and inspection are linked together, as represented in a circular diagram or wheel, which Dr. Deming called the Shewhart Cycle. They are linked because it is desirable to know how well the tolerance limits are being satisfied by the existing process and what improvements are necessary.

Shewhart likened this system, which he called the Scientific Method, to the dynamic process of acquiring knowledge, which is similar to an experiment. Step 1 is formulating the hypothesis. Step 2 is carrying out the experiment. Step 3 is testing the hypothesis. In the Shewhart Wheel, the successful completion of each interlocking component leads to a cycle of continuous improvement. Years later, Dr. Deming popularized and expanded on his mentor's system of improvement in what became known as the Deming Wheel. The wheel symbolizes the transformation toward quality as an iterative process, progressing through each of his four steps: (1) plan, (2) do, (3) check (which he later changed to study), and (4) act. After completing these steps, Dr. Deming taught, the system is modified based upon the results of the analysis. By following this process again and again, knowledge is gained with each turn of the "wheel."

The wheel of improvement symbolized Dr. Deming's relationship with his mentor. He was to take the master's teachings and go beyond, to a new and expanded level. This was the hallmark of his approach—to go beyond. The Deming Wheel and his famous Red Bead Experiment are but two examples.

In the early 1980s, another scientist arrived on the scene and took Dr. Deming's systems theory to a new level. His name was Myron

Tribus. After seeing the NBC special "If Japan Can...Why Can't We?" in 1980, he became obsessed with finding out more about W. Edwards Deming and his teachings. While at MIT, he had an opportunity to interview Dr. Deming, and the experience so moved him that he became a convert to the new philosophy overnight. He was especially interested in the system view of the enterprise, and he and I spent quite a bit of time discussing what became known as the Three Systems of Total Quality. He began writing about Dr. Deming, and one of his earlier works, which I have come to like so much, was called "Deming's Way."

I first met Myron in the mid-1980s, when I was at Florida Power and Light and he was helping to set up a community council in West Palm Beach, Florida. We became friends, and over the years I have come to respect him as a man of boundless energy and vision, a thinker and a philosopher, as well as a master instructor second to none—one who is destined to take the master's teachings to a new and expanded level.

By His Works Shall Ye Know Him

by Myron Tribus

When the history books are written as this century draws to a close, one man will be remembered as having done the most to change the nature of work in this century. Henry Ford and Frederick Winslow Taylor made enormous contributions to factory production. W. Edwards Deming, however, has gone beyond them and has influenced every facet of work, in every industry, in government, in schools, in hospitals...in fact, in every place where the system of work is accomplished through the use of human beings.

Dr. Myron Tribus is President of Exergy Corporation and has held chairs at the Massachusetts Institute of Technology and University of Southern California, as well as the post of Undersecretary of Commerce during the Reagan administration. A well-known quality consultant, he has published hundreds of articles and papers.

Those of us who are engineers, working in industry, know better than anyone else the profound effect Dr. W. Edwards Deming has had on how we think, how we behave, how we see ourselves, and how we relate to our customers, to one another, and to society. Over a full lifetime, Dr. Deming has labored to lift us to a higher plane of service and fulfillment.

There are many reasons to honor the memory of Dr. Deming. We honor and remember him for what he has done, as can be told from the written record. We honor and remember him for what his works signify to us and to all of humankind. And we honor and remember him for what he has become, as he lived his most extraordinarily productive and influential life.

Because all in the quality community share with him a common background in science, systems, technology, and engineering, let us begin our examination of his works by looking at his list of publications. Thanks to Ceil Kilian, his secretary of over thirty-five years, we have his publications list covering the years 1928 to 1991. To what did he pay attention?

In all things he was known as a scholar and a gentleman

In 1928, we find a twenty-eight-year-old newly minted Ph.D. in physics writing about equipotential surfaces for electrons and their effect on the structure of materials. He started his career already at the frontiers of physics. In the same year, we find that he has teamed up with a certain Lola Shupe to write about the mysteries of the thermodynamics of mixtures. It was a good paper. I know, because the same subject continues to be part of my work even today.

Over the next four years, he published a number of papers on the properties of industrially important gases. If you look closely, you will see that most of these papers were published with the same Lola Shupe. In 1934, he concludes his work on the properties of gases and at the same time brings to conclusion another project, which is not documented in the literature but can be inferred from his publications. We see that his co-author has changed her name to Lola S. Deming! We can only guess what was going on during those four years when the record merely shows that he devoted his energies to research.

He did not publish much about this behind-the-scenes project, probably because it did not come under the heading of original research, but the record shows that it produced two daughters, one adopted daughter, five grandchildren, and three great-grandchildren.

In 1934, Dr. Deming began to move away from physics and physical chemistry and published his first paper in the field of statistics. Anyone who has worked on the problem of predicting the properties of gaseous mixtures, based on a study of available experimental data, will not be surprised that this first paper in statistics was concerned with the application of least squares. When you are actually involved in measuring things and comparing what you have done with what other people have done, you begin to understand the conclusion to which he was coming and which he shared with us countless times: "There is no true value of anything." By 1937, he had teamed up with Raymond Birge, known to all scientists for his work on the fundamental constants, to write a paper on the statistical theory of errors. This was no simple recitation of old ideas. The paper ran forty-two pages in the *Reviews of Modern Physics*.

Dr. Deming teamed up with Edward Teller in 1940 to write a paper dealing with absorption of gases. Up to the age forty, therefore, we can think of him as a physicist, capable of working with the leaders in his field, with a strong leaning toward statistics and its use in the interpretation of data.

In the 1930s, Dr. Deming was still deeply immersed in problems of physics, and his publications up to 1940 reflect this interest. We can, however, see a developing interest in statistics. For example, in 1934 he discussed least squares, in 1935 he discussed the chi-square test, and in 1939 he discussed the frequency interpretation of inverse probability. By law, the federal government is required to take a population census every ten years, and in 1940 he became involved with the Census Bureau of the Department of Commerce. His technical background made it natural for him to look at problems of the census in a scientific way. The proper tool for this task is statistics, and so we find in his list of publications a series of twenty-six papers that deal almost solely with problems of sampling, mostly with respect to the problem of taking a census. He stopped his publications in physics about this time.

One paper, published in 1944, during World War II, in mechanical engineering, introduces Shewhart's system of quality control to engineers. This paper reflects two aspects of his career: (1) his work at the Bell Laboratories with Shewhart and (2) his concern that American engineers become acquainted with this powerful systems approach to production quality. We also know, from other sources, that Dr. Deming took the lead in getting this subject into the wartime training of engineers, by giving the first course himself at Stanford University. This is the first level of documentation of the kind of man he was to become.

When there is something that needs to be done, he just goes ahead and does it

In the period up to 1953, we find in his publication list a wide variety of interesting problems to which he applied statistical methods. Among these are:

- Sampling methods in population census
- Errors in card punching
- Training in sampling
- Sampling of the population in Greece
- Judging the quality of market sampling
- Judging the processes of an election in Greece
- Estimating birth and death rates
- Sampling physical materials
- Statistical techniques as a national resource
- Variation in accident rates from automobiles
- Statistics and international trade
- Statistics in legal evidence

These papers were published in several different countries and in several different professional journals. They went far to establish his reputation as a skilled statistician. It would appear that from around 1945 onward, people did not think of him as a physicist but rather as a statistician. It is not surprising, therefore, that in 1948, when General MacArthur needed to make a population survey in Japan, he called upon SCAP to find a replacement for Homer Sarasohn. The search was on and the baton was passed to W. Edwards Deming.

In his publications we find no reference to the Japanese experience with census taking. We do, however, find that in 1953, three years after he started to work with Japanese managers, he began his crusade to bring quality management principles to American managers. In 1953, he published "Management's Responsibility for the Use of Statistical Techniques in Industry," thus marking the start of a theme that he would pursue for the next forty years. He had begun to see the transformation in Japan and tried to warn American managers. However, as Homer Sarasohn points out in Chapter 7, no one listened—at least not in the United States.

From the 1950s to 1980, he continued to make contributions to a wide variety of fields, including:

- Problems of the deaf
- The institutionalization of the elderly
- Census taking with roving populations
- Treatment of patients with schizophrenia
- Newspaper subscriptions
- Teaching statistical methods to people in industry
- The professional responsibilities of statisticians
- Analyzing accounts receivable in motor freight
- Rational costing of motor freight

Another theme begins to emerge. Dr. Deming is a man with a deep concern for humanity, the system of humanity, and the people involved. He reaches out to those who are ill treated in our society. He seeks ways to help them, not by giving direct aid, but by improving the system in which they are caught.

Another characteristic is now apparent in his writings and literature. He cares what happens in society, and when he sees something that needs to be done, he goes ahead and does it, inviting others to join in the necessary work. We who are members of the "quality engineering profession" are aware of how many of our colleagues are indifferent to issues of professionalism. Too many prefer to be simple journeymen, doing only what they must, with little thought of the profession. As is told by the record of his publications, however, the professional obligations of statisticians was heavy on W. Edwards Deming's mind. He wrote a number of papers in which he set forth the obligations of statisticians and, like many of us, had a growing concern that their failure to live up to their obligations deprived society of the benefits of statistics.

I have purposely omitted from this review of the written record his extraordinary activities in the field of management. His books, his seminars, his videotapes, his lectures, and his publications are now so well known there is no need to chronicle them here. His writings have spawned a whole new industry of writing about quality, dominated by people who, having almost understood what he has said, scurry into print to explain it to the rest of us.

Prior to going to Japan, it is evident, as we read about his experiences with Shewhart, that Dr. Deming began early on to see what could be done by enlightened management. He observed the way people were treated under what was called Taylorism or Taylor's Scientific Management, and he knew something of the costs in human, financial, and physical terms. There was waste all around, and he saw, more clearly perhaps than Shewhart himself, how all of this could be traced back to management. His attempts to interest American managers came to naught, for we were then in an era of unprecedented growth and expansion, in which any damned fool could make a living, and the more ruthless could make a fortune.

His impact in Japan has now been told many times, and I shall not discuss it here for it is covered elsewhere in this book. What needs to be said, however, is what these teachings, which can be summarized in the following ten management actions which have been adapted from his new systems philosophy, reveal about the man himself:

1. Recognize continuous improvement as a system.
2. Define it so others can recognize it, too.
3. Analyze its behavior.
4. Work with subordinates in improving the system.
5. Measure the quality of the system.
6. Develop improvements in the quality of the system.
7. Measure the gains in quality, and link these to customer delight and quality improvement.
8. Take steps to guarantee holding the gains.
9. Attempt to replicate the improvements into other areas of the system.

10. Tell others about the lessons learned, leading us to the ultimate benefit of mankind.

He has a deep and abiding love for the human race, and he saw that managerial practices were causing human waste and suffering. He dedicated his life to rectifying the root causes of our misery. He saw what needed to be done, and he set about doing it.

He knew that the only salvation for the Japanese would lie in a transformation of their management, and so he insisted on lecturing to them. Why did he succeed? Well, for one thing, the Japanese were desperate, and they listened. But they were also influenced by the fact that he was, and is, a generous, kind, and loving man. They say so in their recollections of his lectures. This side of Dr. Deming is difficult for those American managers who have attended his lectures to understand, for when he sees them behaving badly, he does not hesitate to scold them publicly. They see him as tough and irascible. While he is tough on managers, he is gentle and caring with workers. If you do not believe this, take a look at the videotape of his discussions with workers at the Pontiac plant, made in 1981.

In Japan, another side of his character is evident in his refusal to take any royalties from the Japanese translation of his lectures. Instead, he donated them to start what became known as the Deming Prize. While his generous nature is renowned in Japan, it is little known in the United States, but I have seen it in gifts he has made to teachers and authors—with no fanfare or public relations.

He saw what needed to be done and he set about doing it

A turning point in his career came in 1980, when NBC broadcast the famous special "If Japan Can...Why Can't We?" Just before the program aired, both Ford and Pontiac had sent teams to investigate what was going on in Japanese auto manufacture. The teams came back mystified. They saw nothing different—the same equipment, the same flow of materials, nothing spectacular in the way of automation. As the two teams puzzled over what they had seen, they also happened to see this now-famous program, in which there were a few shots of Dr. Deming. At one point, he said of American manufacturers:

> Inspection does not build quality; the quality is already made before you inspect it. It's far better to make it right in the first place. Statistical methods help you to make it right in the first place so that you don't need to test it. You don't get ahead by making product and then separating the good from the bad, because it's wasteful. It wastes time of men, who are paid wages; it wastes time of machines, if there are machines; it wastes materials.

It so happens that several months later, I personally visited both Ford and Pontiac and met with some of the team members who had gone to Japan. They confessed that without the insights they got from Dr. Deming's brief appearance on that NBC special, they would not have understood what they had seen. This was the beginning of the transformation in America—late, of course, but nevertheless a beginning.

In the early 1980s, Dr. Deming often said, "I lit many fires, but they all went out." Today he can say that no more, because there are people all around the globe who study his books, watch his videotapes, and follow his teachings in an almost religious way.

No one knows how many people he has touched either directly or indirectly. It must be a number measured in the millions. First of all, there is the entire Japanese population. During one trip to Japan, we saw his teachings at all levels, from high-tech factories to clerks working in a tourist center. We even saw a young woman working in the Bunny Club of Kyoto who was applying quality management tools to the operations of the club. In France, there is the French Deming Association. In the United Kingdom, there is the British Deming Association. In Australia and New Zealand, there is the Total Quality Management Institute (although Dr. Deming does not believe in the phrase "Total Quality Management"). Across the United States, there are now about 300 local groups, called Deming User Groups, devoted to the application of his ideas in their local communities.

From his works, therefore, we know a great deal about the man. We know what he has become in the eyes of millions of people: a man of great wisdom, dedication, and selflessness, determined, so long as there was breath in his body, to save the human race from its inherited follies.

I ask you to join me in paying respect to the memory of W. Edwards Deming—a man who saw what needed to be done and just went ahead and did it. He saw the need for a superior system of management. Single-handedly, he created it, taught it, and, through tireless lecturing, writing, and campaigning, made it a way of life for millions of people.

We, as well as succeeding generations to come, owe him much more than we can ever pay. Let this work begin to signify our acknowledgment of that indebtedness, which can never be repaid.

It seems to me that the prime requirement for a teacher is to possess some knowledge to teach. He who does no research possesses no knowledge and has nothing to teach. Of course, some people that do good research are also good teachers. [However] the student is at a disadvantage when asked to evaluate his teacher. On what basis? Luster of personality? Knowledge of the subject? Content of the course? Is he communicating to the student what he is trying to say? The only suitable judge of a teacher's knowledge are his peers. The only objective criterion of knowledge is research worthy of publication—measured on some scale of contribution to knowledge, not by numbers of papers.

W. Edwards Deming
The American Statistician
Vol. 26 No. 1, February 1972

Chapter Four

INSTITUTE TRAINING AND RETRAINING

"Institute training and retraining on the job."

Dr. Deming believed in the need for a total transformation of the training philosophy that is followed in most organizations. Time and time again, he found new workers being assigned to production jobs with little or no training in the proper techniques by co-workers or supervisors. He believed that training in basic statistical tools and techniques would be useful to most workers and would complement normal job-skills training, in addition to allowing the individual to make a greater contribution to quality improvement.

He also taught that retraining is an important component in keeping employees current with the latest job skills. Retraining workers who are displaced by automation or downsizing should be a standard practice and would serve to build loyalty between employees and their company. He emphasized that education and training efforts point everyone in the organization toward one goal: producing the best possible product or service to satisfy the customer. Education and training are investments, not expenses, because they equip people to make solid improvement decisions based upon incisive analysis of data.

A recent study of the training and development practices of U.S. businesses revealed that the average American company spends about one-half of one percent of its annual payroll budget on training. In contrast, the excellent organizations, such as Motorola, Xerox, Hospital Corporation of America, and others, invest about five percent in training, which translates to between ten and fifteen classroom days of training per year—ten times the national average.

The stories you are about to read are by May Lum Gould and Jay Gould. They are one of the many husband and wife teams who have embraced the Deming philosophy. May Lum began her quality odyssey with the extensive training she received from an American company that was to transform the culture and system of production in Malaysia. Jay Gould helps fill in the gaps and gives us some insight into Dr. Deming's impact on the military establishment.

My First Quality Guru: Dr. W. Edwards Deming

by May Lum Gould

On our first blind date, my husband, Jay, and I spent the majority of the evening discussing the quality program that U.S. Motorola personnel taught in their Malaysian plant. I was the group team leader for production improvement, and it was my team's task to improve the quality of our products.

Jay asked me if I had ever heard of Dr. Deming. "No," I said, "who is he?" He laughed and said, "Never mind, we can talk about that later. First, I wanted to know something. Tell me what you have learned about American quality. Would it work in a marriage?" I knew it would and told him so. Everyone knew that in America, people did their best to make things better in their work and in their homes. My belief in American quality impressed this man, who in

---

May Lum Gould has been a follower of Dr. Deming since the early 1970s. As a Motorola worker in Malaysia, she learned firsthand what total quality was all about.

seven short days became my husband. Talk about shortening the cycle time!

Shortly after our marriage, he told me about Dr. Deming and his work in Japan. I read parts of all the books in my husband's quality library. I studied Dr. Deming's Fourteen Points and began to see why American quality was so great. I guess I was very naive about America, because as my husband spoke more and more about Dr. Deming and the American quality movement, I began to wonder about all those things being taught by the Americans at the Motorola plant in Malaysia.

When Motorola arrived in Malaysia over twenty years ago, the trainers presented some "American theories" about how the people in the plant should work and get along. They explained, "We do not allow discrimination. In America, we all work as a team. It does not make any difference if you are a man or a woman. It does not matter if you believe in a different religion, if you are of a different race, or if you speak a different native language. At Motorola, you are part of a team and not Malay, Chinese, or Indian. We all have the same uniform, from top management down to the production operators. Everyone will speak English or Malay and be trained as team members. You and your team will receive a bonus for your good, hard teamwork that produces good quality products with zero defects. Doing this together, we will all make more money than you have ever made before. And the lunch will be on us, the company."

Well, we all wanted to be like the Americans and be good to each other. If one team member fell behind, we would say, "You have to catch up. Americans would never be like this. We must find a way to improve, as the Americans always do." Everyone knew that the Americans made the finest products in the world, and we all wanted to be like them. The American hero was a very real role model for us.

Motorola taught us many great things: what leadership is, how to work together as a team to produce quality products, how to improve the quality and reliability of our products, how employers and employees should maintain a harmonious relationship, and, best of all, it helped us improve our quality of life. All this served to reinforce the model. American firms investing in Malaysia gave Malaysians the opportunity to earn a better living.

The interesting thing is that we did not realize that what the Americans taught us was an early form of Dr. Deming's work in total quality (TQ). The part that we liked the most was the eighth of his Fourteen Points: drive out fear. Dr. Deming said:

> It is necessary for better quality and productivity that people feel secure. *Se* comes from Latin, meaning without; *cure* means fear. *Secure* means without fear—not afraid to express ideas, not afraid to ask questions.

In Asian countries, the majority of workers labor under conditions that are intimidating. Motorola had an open-door policy where everyone was welcome to walk into the managing director's office to talk—at any time! Motorola wanted us to understand that we are truly working as a team, from top management down to the production floor. By doing so, they made us feel appreciated. We were very happy to work for the company.

One day my husband came home and said, "You are going to meet Dr. Deming. He is giving a seminar down on the coast, and I have an invitation to assist him." My husband worked primarily with Dr. Joyce Orisini at the time, who was one of Dr. Deming's closest assistants. She is a pleasant and bright lady whom I liked immediately. And then there was Dr. Deming—very gruff, at times not at all friendly. My husband called him "the old curmudgeon." Maybe to some he was, but he was always polite and kind to me.

When it came to discussing quality, we were confident that Dr. Deming knew it all. He was always very tough on corporate management and believed that management did not trust the workers. In fact, management blamed the workers for all the problems that, according to Dr. Deming, were due to the system, which, as he taught, is owned by management. He believed that the job of management is not supervision, but instead leadership and taking responsibility for the job of fixing the system.

At Motorola, we saw firsthand how engineers work closely with production operators. Management understood that the operators knew their jobs well. After all, management had trained them using the American training methods. The engineers needed feedback from the operators to improve their work and in turn were eventually able to help improve the operators' work. People helping solve one another's problems seemed very synergistic to us at the time.

That first meeting with Dr. Deming was very memorable. George Washington University even gave me a certificate for attending the five-day lecture. (Most of all, I think they gave it to me for being at Dr. Deming's side when he needed a helping hand.) When we returned home, I found a new interest in my husband's work. He was the Deputy for Total Quality at the Ballistic Missile Office, and his job was to bring the TQ culture into the Air Force Command. Although it was a shock to me, he explained how the Defense Department had rejected the work of Dr. Deming after Word War II as unnecessary in a peacetime economy. Statistical controls were apparently only an emergency measure, and after the war, the secret classification for quality control was downgraded. However, the Japanese learned of the "secret" through Dr. Deming, and the rest is history.

As time went on, I began to read and study more of my husband's books, a little at a time, trying to understand what this new philosophy was all about. More and more it seemed to me that Americans are not the perfect role model that we in Malaysia believed them to be; they are just like the rest of the world. The irony is that because of the ideas brought to Malaysia by the Americans, we, like the Japanese, are now ahead of the United States in many areas.

Many books have been written about Dr. Deming's work. One of my favorites is by William W. Scherkenbach, who did his doctoral studies under Dr. Deming and Dr. Orisini. When Dr. Deming received a call for assistance from Ford CEO Donald Peterson, he flew to Detroit with Dr. Orisini and Mr. Scherkenbach. Scherkenbach found a home in Detroit and stayed on to guide Ford's quality journey. His book is about his adventure at Ford and how he used his experiences there to explain Dr. Deming's Fourteen Points.

Mary Walton is a Philadelphia newspaper reporter who wrote a feature article in the *Philadelphia Inquirer* about Dr. Deming's work which eventually resulted in the publication of a couple of books. Mary's books convey the story of how Dr. Deming has changed the course of not only corporate bodies in the United States, but cities and communities as well. Again, there is strong emphasis on the Fourteen Points, but she also includes the Seven Deadly Diseases, one of which is the cost of health care, which is seen by Dr. Deming as an early villain. How prophetic Dr. Deming was, and how brilliant are those like Mary Walton who wrote about him. Her works are like a primer on how to change to the Deming quality method. They bring to life Deming's Red Bead Experiment and dramatically present the Funnel Experiment. These experiments are analogies or parables about corporate life and producing a better product for less.

Mary quoted the Deming way of using statistical control charts, which are charts used to analyze processes. The purpose, Dr. Deming emphasized, is "to stop people from chasing down causes." Properly understood, a control chart is a continuing guide to constant improvement. She also told of Dr. Deming's influence on the Japanese gurus of the day. Dr. Kaoru Ishikawa wrote a text on the types of controls used in the metrics of the Deming way. He developed techniques used to pinpoint the source of variations that cause product deficiency. This was the same Dr. Ishikawa who developed the first cause-and-effect diagram in 1943, which has proved so useful in any problem-solving situation. Dr. Deming felt that a cause generally consisted of many complex elements. Therefore, his diagrams usually turned out to be rather complicated, especially to those of us who were just learning the system. Dr. Masaaki Imai presented the Japanese view of continuous improvement, which was known at the time as the Deming Wheel, or the P-D-C-A Cycle. He put together management philosophies, theories, and tools developed and used over the years in Japan and organized them into the concept known as *kaizen*. He taught that this concept is so natural and obvious to many Japanese managers that they often do not recognize it as a concept. According to the *kaizen* philosophy, one's way of life deserves to be constantly improved.

It is ironic that both of these Japanese men were the first students of Dr. Deming in Japan. U.S. Army Sergeant Homer Sarasohn had taught both of them statistical theory and the World War II "secret" before Dr. Deming went to Japan. Under Dr. Deming, they continued to learn more about the applications, and they added ideas of their own.

Dr. Deming has been to our home and we to his. At my husband's invitation, Dr. Deming came to his campus in January 1992 to give a seminar. He did it as a friend for a friend. During the seminar he lectured and did what he normally does during one of his presentations—he chewed out the generals in attendance for their lack of management understanding. Then he lectured them on the Funnel Experiment and graciously allowed the audience to participate in his Red Bead Experiment for the thousandth time.

Now, a year later, Dr. Deming is no longer with us. Prostate cancer had spread to his lungs and bones, and toward the end he suffered oxygen deprivation and occasionally passed out. But his belief that America can turn itself around remained strong to the very end. He somehow got up out of his hospital bed each week to fly, in a hospital plane and with attending nurses, to his conferences at Ford or seminars around the country. He believed that his work was vital to America regaining its place in the world market. He will not rest in peace until this is accomplished. My husband and I miss him greatly and mourn his loss.

It seems so long ago when my husband first heard about Dr. Deming, when Jay was at West Point. Since the 1980s, Dr. Deming has been the principal cause for change in both of our lives. We cherish the memory of his friendship and will hold his memory dear. There are many in the Deming camp, as well as those who have their own camps, who are good friends to my husband and me. All of us have an overriding goal: to see America great again.

We said our good-byes to Dr. Deming a while back, on September 22, 1993. We bought his final book, which he signed, and we bid our farewells. I spent the next few days in deep reading, wanting to get his last thoughts into my mind before he passed away. He described the theory of a system and introduced the system of Profound Knowledge in this book. The system of Profound Knowledge provides a lens with which to understand and optimize an organization so that the work would benefit everyone—stockholders, suppliers, employees, and customers. We will all miss him. May God be with him at the moment of his rest.

The work of Dr. Deming, which I learned in the beginning from my teachers at Motorola, has helped me in many ways. I am able to understand how to work within a team and how continuous improvement is a very important part of the system, not only in the workplace but in whatever we do. My husband and I have truly applied what we have learned in our marriage, and it is amazing how wonderful things have been for both of us, even when we encountered problems that seemed impossible to manage. Motorola has been very successful in Southeast Asia as well as in America and provides a good example of how concrete American theories really work. It is an honor to know so many of the great American quality philosophers. To have known the greatest of them all—Dr. Deming—is the greatest honor of all.

Bringing the Deming Philosophy Home

by Jay Gould

I moved to California to accept a new position and make my contribution to the deployment and activation of the Peacekeeper Weapon System. I gained the technical knowledge in the Titan I and II days of the early U.S. efforts to deter war with the Soviets by constructing retaliatory weapons. It felt good to be back working with the people I understood and enjoyed. We ran the deployment and activation part of the program based on the same quality principles I had learned in the early 1950s. The effort was under a great deal of micro management from Congress, and I found myself very busy just staying ahead of the game. As the pressure decreased and things began to fall into place, I found myself very lonely. Three things came into my life which changed things forever. First, I decided to get my master's degree at the University of Southern California. Second, I made the personal acquaintance of a man I had heard about for years and years, Dr. W. Edwards Deming. Third, I met my future wife, May Lun Lum, on a blind date.

In each course at USC, the writing of a course paper was a requirement. In one particular course, the class was divided into teams. Our team elected to write on the worldwide quality movement emanating from Japan and precipitated by Dr. W. Edwards Deming. With a high-performance work team dedicated to task accomplishment, wonders can be performed—and we did. The paper we produced gave us all a new insight into what we believed was a wave of change about to sweep across America.

At that time, the U.S. Air Force Ballistic Missile Division sponsored a seminar given by Dr. William W. Scherkenbach. At that seminar, I learned that Dr. Deming would be in the Los Angles area the following week with his four-day seminar. I was overjoyed. I had learned of Dr. Deming's "secret" work while at West Point. Bill

---

Jay Gould has been a follower of Dr. Deming since the 1950s, when he was a student at the University of Southern California working toward his master's degree. His career in the Air Force gave him a military perspective toward total quality.

Scherkenbach, a Naval Academy graduate who took his doctorate from Dr. Deming, assured me that Dr. Deming would enjoy meeting me. When I reported for duty the next day, I immediately filed my leave papers and made arrangements to travel to Newport Beach. When the commanding general heard of my resolve, he immediately notified me to withdraw my leave papers because he had signed official orders. It would be in the best interests of the Air Force if I were to officially attend Dr. Deming's seminar and learn whatever I could. At the seminar, Dr. Deming most graciously received me and made arrangements for me to work with him when he came to the West Coast.

Then God blessed my life. I met May Lun Lum on a blind date. I received a letter from her, which I found to be the most gracious and sincere letter I had ever received in my life. I bought a dozen roses and drove to pick her up. Traffic was light, and I arrived very early but decided to ring the bell anyway. When she emerged from the house, I knew at first sight that she was a person of quality. She apologized for not being ready, and I felt a bit of a fool. I gave her the flowers and suggested I would come back in an hour. I raced to the local Chinese jewelry store and bought a gold bangle.

She impressed me with her pleasant personality. As we drove to dinner, she informed me that she was in the United States on vacation from her position as a production group leader in the Penang Motorola Plant. Her team's tasks were not only to produce quality products, but also to learn new concepts and techniques to help improve quality. Everything fell into place. I knew of Motorola's quality effort from my research on the USC team paper. I knew of her task because of my association with Dr. Deming. I knew I was talking to a kindred soul. We spent the whole date talking about quality in the workplace and how it comes about.

The date was marvelous, and as I took her to the door I asked her out again the next week. Her response startled me. "I think that would be great," she said jokingly. "What time do you think your plane will land? My visa is up and I must return home." The response really shattered me. I stood on the brink of a decision. I had never met anyone as charming and intelligent, and now she was about to get on a plane and fly away. I could not let that happen. From somewhere I heard my voice saying, "May, do you think the quality we have discussed tonight could happen in a marriage? Would you be willing to see me every night this week so that we could discuss possibilities?" With a rather questioning and yet serious look in her eyes, she responded, "Yes, let's talk."

One week later, much to the surprise of my fellow workers, we were married by the justice of the peace in San Bernardino. My division commander and his intended stood up for us. My USC quality team members bought the cake. I carried May across the threshold of our home and said, "I give you this house. Do with it as you will for it is ours." She responded, "You do me great honor in being my husband." The next day, when I returned from work, the whole kitchen was different and looked beautiful. I thought, "She must have spent a feverish day working on it." Out loud I exclaimed, "Oh! How beautiful. You have outdone yourself." I spent the next half hour explaining what outdone meant.

When I arrived home on the second day, the same thing occurred. The kitchen looked different again. My private thoughts were, "It looks better, but yesterday was just fine. I wonder why she did it again?" I told her how good the changes looked. Well, this same thing went on for two weeks. Every day the kitchen looked different, and it bewildered me. There is not a great deal you can say to a lady about such a thing when you have only known her for three weeks, two of which you have been married to her. Finally, it got the best of me and I blurted out, "May, what is happening here? Each day the kitchen is different." May responded, with that wicked smile in her eyes, "Jay, you should know better. This is the process of continuous improvement. It will take me six months to make this into a good quality Chinese home. Now, *you* are going to take a little bit longer."

Our home has truly become a quality Chinese as well as American home. More often than not, we share our home with our fellow workers to enjoy a wonderful "Chinese eight-course meal." Dr. Deming was a frequent visitor, as are Dr. Joyce Orisini, Dr. Myron Tribus, Mr. David Langford, Dr. Cathy Kitchen, and others of the quality movement who have been here to share in our joy. As May pursues her degree in computer information systems, I work on my doctorate. By the way, May was correct as usual—I did take a little longer.

Everyone might well ask himself every day what he has done this day to advance his learning and skill on this job, and how he has advanced his education for greater satisfaction in life.

W. Edwards Deming
Out of the Crisis

Chapter Five

INSTITUTE MODERN METHODS OF TRAINING, EDUCATION, AND SELF-IMPROVEMENT

"America must begin to institute modern methods of training and a vigorous program of education and self-improvement."

Dr. Deming often told stories about companies impacted by technology in such a way that vast numbers of people needed to be shifted into different types of work or just plain let go. Although technology may add some jobs in some areas, many others will disappear. He cautioned that quality must not be perceived as costing jobs, or morale and teamwork will suffer and the workers will lose heart. He taught that it takes management with long-range vision to commit to this kind of self-improvement program and make it work over the long haul.

He would often say that having good people is not enough. They must continually acquire new skills and knowledge in order to learn new methods of production. He felt that people had a general fear of education.

> People are afraid to take a course. It may not be the right one. So what! Take it anyway and find the right one later. For you do not know what is the right or wrong one. Study, learn, improve and find the right one later, for you never know what is practical and can be used later on.

To illustrate this point, he told the story of Kansai Electric, the first electric utility in the world to win the Deming Prize. It seems that Kansai Electric developed total quality programs for all 20,000 of its employees, starting with courses for management and expanding to every worker in the form of general employee courses for education and self-development. These courses were coupled with manuals and case studies as a way of extending this self-improvement to all areas of quality. Kansai Electric did not stop there, however, he noted. They also made this training available to their suppliers on an as-needed basis. Dr. Deming called this "influencing the extended enterprise." They were instituting modern methods of training and self-improvement for everyone involved.

Dr. Nancy Mann has been a disciple and follower of Dr. Deming since the late 1960s. As a mathematician with a Ph.D. in biostatistics, she has been heavily involved in the American Statistical Association, an organization in which Dr. Deming was also an active member. For the last several years of his life, Dr. Mann's organization, Quality Enhancement Seminars, Inc., was the producer of all of Dr. Deming's public seminars. Her book, *The Keys to Excellence: The Story of the Deming Philosophy,* contains outstanding insights and quotes from Dr. Deming, which helped to fill in many of the holes in the history of the movement. The following tribute reflects the key ideas that she puts forward in her book, which clearly show the influence of the modern methods.

Dr. Deming: Holder of the Keys To Excellence

by Nancy Mann

"We are in a new economic age. We can no longer live with commonly accepted levels of delays, mistakes, defective materials and defective workmanship."

I first heard these words from W. Edwards Deming a little over a year after I met this remarkable man. At that time, I was developing statistical methods for the analysis of reliability data at the Research Division of Rocketdyne, where rocket engines were being produced for the Apollo moon program.

Dr. Deming was then a consultant for several large corporations, which included carriers of motor freight, railroads, telephone companies, and the like. In this capacity, he had attained some prominence as an expert court witness. In addition, he was a professor in the Graduate School of Business Administration at New York University. It was both his expert witness status and his NYU affiliation that led to our first conversation.

Early in 1968, I was helping to put together a conference entitled "Computer Science and Statistics: A Symposium on the Interface," for which the organizing committee had decided to have a session on jurimetrics, which is statistical aspects of court proceedings or the treatment of statistical problems in court. We invited Dr. Deming to come to Los Angeles and give a jurimetrics paper at our meeting. He accepted and arranged to adjust his schedule so that his visit would follow a consulting trip to San Francisco. As a result, there would be little or no cost to the conference for his travel expenses.

All was well until we sent Dr. Deming a copy of the conference brochure, in which he was listed as "W. Edwards Deming, New York

---

Dr. Nancy Mann is a member of the International Statistical Institute and a Fellow of the American Statistical Association. Her doctorate in biostatistics is from UCLA. She traveled extensively with Dr. Deming, and her organization, Quality Education Seminars, conducted his famous four-day seminars.

University." Upon reading this, he phoned to say that he liked to be given credit as someone who was able to make a good living as a "Consultant in Statistical Studies" and that he therefore had decided not to come to our meeting and give the talk.

The job of calling him to apologize fell to me for two reasons. First, I was the one who was responsible for the nearly fatal error. Second, I was chairman of the conference and the buck stopped with me. As it turned out, only a few seconds of apology were necessary; he conferred instant absolution as soon as I claimed responsibility for the gaffe. He came to Los Angeles as planned and helped to make the conference a resounding success.

We met again in May of 1969 in Washington, where I had gone for a conference at which I was invited to speak. Some time before the conference, I received an invitation to join Dr. Deming and another conference speaker for dinner at the Cosmos Club on the first evening of the event. I accepted, and upon my arrival in town, I touched base by phone to find out when I should meet them for the occasion. It was then that I discovered that the other speaker was unable to join us.

That left just the two of us to eat and talk, after we met in the ladies' parlor, just inside the ladies' entrance to the Cosmos Club. Mealtime provided a chance for me to find out how Dr. Deming, who was originally trained in physics and mathematics, had made such an impact on the discipline of statistical quality control and had come to have so much influence in its being applied in the United States and in Japan.

The story should begin in March of 1938, ten years after Dr. Deming earned his Ph.D. in physics at Yale and shortly before he left his position as a mathematical physicist at the U.S. Department of Agriculture (USDA) to join the U.S. Bureau of the Census. At the time, he arranged for Dr. Walter Shewhart, the father of process control, to deliver a series of four lectures entitled "Statistical Method from the Viewpoint of Quality Control" at the USDA Graduate School, where Dr. Deming had charge of courses in mathematics and statistics. Dr. Shewhart was based in New York City at the Bell Telephone Laboratories, where he and Dr. Deming had met some years earlier. Working together, the two became fast friends. Dr. Shewhart's lectures at the USDA were published by the Graduate School in 1939 "with the editorial assistance of W. Edwards Deming."

When Dr. Deming began work at the US. Bureau of the Census,

it occurred to him that quality control procedures could be applied to the routine clerical operations of the 1940 population census, such as coding and card-punching. During the learning period, the error rate of an individual punching cards was high, but with training and experience, a good card puncher's error rate dropped markedly and could be brought under statistical control. At first, the work of all card punchers received 100 percent verification or correction; later nearly 40 percent qualified for only sample verification.

Work subject only to sample proofing flowed through the process six times faster than otherwise. In an article in the September 1941 issue of the *Journal of the American Statistical Association,* Dr. Deming and Leon Geoffrey estimated that the introduction of quality control saved the bureau several hundred thousand dollars, which was transferred to other work in addition to, above all, contributing to earlier publication of census results.

Dr. Deming's seven years at the Bureau of the Census began in 1939. Users of census data have always wanted more information than can possibly be provided with a normal budget, and in the late 1930s the demands were larger than ever before. Many of the users of census data were willing to accept sample results, but some of the old-timers at the Bureau of the Census were opposed to the idea of sampling. "Sampling was abhorred," Dr. Deming told me, "because the census had always been complete. It couldn't be anything other than complete. But sampling was in the air."

The final decision rested with Secretary of Commerce Harry Hopkins. After listening to the arguments pro and con, Hopkins decided in favor of sampling. This meant, of course, that the bureau would need some expert help in designing the sampling procedure that would be used in the 1940 population census. "Well," Dr. Deming told me, "one day in 1939 the telephone rang, and it was Dr. Philip Hauser, the assistant director of the Census Bureau, wanting to talk with me about a job. And I said, 'Right away!'" Use of statistical quality control procedures has been a standard practice at the Bureau of the Census ever since.

The next relevant series of events began early in 1942, soon after World War II had broken out. Dr. Deming, who was then working at the Bureau of the Census and was also a consultant to the U.S. secretary of war, received a letter from W. Allan Wallis. Wallis, who later became undersecretary of state for economic affairs, was at that time on the statistics faculty of Stanford University and concerned that Stanford seemed to be relatively untouched by events in the world outside the campus. He and several other members of Stanford's statistics faculty were seeking guidance on how they might contribute to the war effort.

> Here is my idea. Time and materials are at a premium, and there is no time to be lost. There is no royal shortcut to producing a highly trained statistician, but I do firmly believe that the most important principles of application can be expounded in a very short time to engineers and others. I have done it and have seen it done. You could accomplish a great deal by holding a school in the Shewhart methods some time in the near future. I would suggest a concentrated effort—a short course followed by a long course. The short course would be a two-day session for executives and industrial people who want to find out some of the main principles and advantages of a statistical program in industry. The long course would extend over a period of weeks. It would be attended by the people who actually intend to use statistical methods on the job. I would suggest that both courses be thrown open to engineers, inspectors, and industrial people with or without mathematical and statistical training...

The project was subsequently undertaken by the War Training Program and was such a success that, beginning early in 1943, intensive eight-day courses in statistical quality control were given at many universities throughout the country under the auspices of the U.S. Office of Education. Dr. Deming taught twenty-three of them. Within two years, almost 2000 men and women from nearly 700 industrial concerns had attended the courses. Many of the students went on to serve as instructors in part-time courses, in which another 31,000 persons in industry participated.

The program had a strongly beneficial effect on the quality and volume of war production. Spectacular reductions in scrap and rework were made. Process control, however, was used mostly as a tool for dealing with local crises. Meanwhile, the wartime experience helped lay the groundwork for the establishment of the American Society for Quality Control (ASQC) in February 1946. Dr. Deming, who played an important role in the founding of this society, told me:

> Wherever I taught, I told the people, "Nothing will happen if you don't keep working together. And you've learned only a little, so you must keep on working and meeting together." They did, and out of that nucleus grew the ASQC.

So it was that methodology for improvement of quality began to evolve in the United States, but it failed to realize its potential. If things had been a bit different, it might have been the United States that experienced a quality renaissance. Conditions were right in many respects, but an essential ingredient was missing, namely management's awareness that there was a growing problem and that there was a means to deal with it.

During his many visits to Japan, Dr. Deming often had opportunities to get together with some of the people whom he had met during his first trips there in 1946 and 1948, such as Dr. Nishibori and Mr. Ken-ichi Koyanagi, for example. On one such occasion, Dr. Nishibori confirmed that it was Dr. Deming's earlier encouragement that had precipitated a suggestion of his about using statistical methods to help with the rebuilding of Japan. One of the principal problems of Japanese industry was that the previously captive markets of China and Korea were lost to them after the war, and they needed to trade so that they could import food.

In late 1949, Dr. Deming received a letter from Mr. Koyanagi, who was by then the executive director of the Union of Japanese Scientists and Engineers (JUSE), asking him to come to Japan to teach statistical methods for industry. Although he could not go immediately because he had too many projects under way, he did finally make the trip in June of 1950 under the auspices of the Supreme Commander of the Allied Powers.

Dr. Deming described to me the fateful events that eventually involved the industrial leaders of Japan in the educational process and provided the critical impetus for changing the image of Japanese products:

> They were wonderful students, but on the first day of the lectures a horrible thought came to me. "Nothing will happen in Japan; my efforts will come to naught unless I talk to top management." By that time I had some idea of what top management must do. There are many tasks that only the top people can perform: consumer research, for example, and work with vendors. I knew that I must

reach top management. Otherwise it would just be another flop, as it had been in the States.

I do everything the hard way. That's one way to do it. I immediately talked to American friends who knew the right Japanese, and before long I was talking to Mr. Ichiro Ishikawa, who had formed the great Kei-dan-ren, the Japanese association of top management. He was also president of JUSE, which I did not know at the time. I knew Mr. Koyanagi; he was managing director and did all the work. The office of president of JUSE was more or less honorary. All of this would have been so simple if I had only been aware of these facts.

After three session with Dr. Deming, Mr. Ishikawa saw what needed to be done. He sent telegrams to the forty-five top industrialists, telling them to come to the Industry Club the following Tuesday at five o'clock to hear Dr. Deming. They all came. Dr. Deming told me:

> I did the best I could. I gave them encouragement. That was the main thing. I told them that the old days are over and that you can produce quality. You have a method for doing it, for you've learned what quality is.

The explanation of why this miracle happened in Japan lies in the coming together of all the factors necessary for a transformation. After the war, Japanese industry experienced a "bottoming out." W. Edwards Deming came upon the scene and learned to appreciate the Japanese personality and culture. He had an awareness of what needed to be done, and he saw to it that the message was communicated to the people with the ability to take action. The Kei-dan-ren, the organization for Japan's top management, supplied the means for getting the message to all of Japanese industry, as well as the means for continued cooperation. JUSE was in place, ready to provide training and facilities. And the Japanese cultural heritage created precisely the national psychology that made product excellence a reality. "Now people in this country are beginning to be aware of the miracle and are trying to find out about it, thinking they can copy the Japanese," Dr. Deming said. "They cannot. People can't go to Japan and learn because they don't know enough about the subject."

Dr. Deming set down fourteen cardinal points encapsulating his philosophy. The wording and format have changed over the years, and communication of the essence of the philosophy is still evolving over time. This evolution was accelerated in the last few years of his life because of the tremendous amount of feedback he received, from both his consulting and from discussions with the several thousand who attended his seminars each year. He was always learning.

When his Fourteen Points first appeared, two fundamental ideas supported them. The first was that management must strive to develop business over the long, the very long, term; it is not enough for a business to make a quick profit. The second was that this goal can be attained only by the delivery of high-quality, dependable products and services.

Few would, in theory, question this second precept. Yet one cannot help but notice that if turning out superior products and services is, in fact, a goal of Western industry, something has been lost in the translation to reality.

We are all busy making mistakes, and most of us agree that is what gives us the greatest potential for learning. The problem seems to be that we are not making use of this potential and profiting from our mistakes. In Dr. Deming's philosophy, statistical data-analysis techniques are used as mechanisms for tapping and exploiting the potential information generated by the processes turning out goods and services: to anticipate, identify, and correct mistakes and to reduce variability in the system in order to improve quality and achieve excellence. Along with improved quality attained by these means are the added dividends of decreased costs through the better use of human effort, machine time, and materials, as well as a decrease in rework and scrap. Productivity is thereby increased, markets are captured and ensured over the long term, and jobs are provided.

The philosophy is built around the Shewhart Cycle. This cycle, which Dr. Deming described to the Japanese in 1950, uses testing and statistical feedback at various stages of manufacturing and marketing in order to constantly improve a system which involves design, production, merchandising, and service. The Deming philosophy, by making use of the Shewhart Cycle, shows us a way of doing business based on the realization that the customer and the level of customer satisfaction are the most critical elements in the entire production process. In the Deming doctrine, customers are those who buy the product as well as those involved in succeeding stages of production.

In addition to working toward customer satisfaction, Dr. Deming early on emphasized the importance of satisfaction of the worker and of management, including the elimination of adversarial positions. The major overall goal is productivity through efficiency and delivery of a product in which workers and management can take pride and which consumers will buy because they believe in it.

Constancy of purpose in the early 1980s was the cornerstone of the Deming philosophy. The use of statistical thinking was of equal importance. Statistics is the essence of the scientific method and relates to almost everything we do. It is a discipline that deals with chance and choice, with tradeoffs, with cause and effect, and with predicting future events from a collection of facts and numbers. Statistics was used, as told in the Bible, when Joseph devised a plan for allocating to the Egyptian people grain collected in "good years," but it was not until the 1920s that statistics began to develop as an organized discipline. This development came about in large part through the efforts of a Cambridge-educated English scholar, Ronald Aylmer Fisher, whose work in designing agricultural experiments and in the application of the theory of genetics led him to develop a theoretical framework for prediction by statistical methods. His thinking struck so many as electrifying that eventually people came from all over the world to study with him at the University of London.

Among those who traveled to London to learn from Fisher was W. Edwards Deming; he had been exposed to a smattering of statistical methodology during his training in engineering, thermodynamics, and astronomy. What Dr. Deming gleaned from Fisher's lectures, together with many later experiences and his fascination with the theory of statistical process control, led to the development of the statistical core of the Deming philosophy. This statistical core has defined a way of life in Japanese industry.

On one occasion, I asked him how he came to be a statistician. His answer was quite involved:

> Well now, that's a very difficult question. I don't know if I have a good answer. But I don't know who else could answer it. Courses in engineering and surveying led me to the theory of errors, and in studying physics and mathematics, I learned a lot of probability. Kinetic theory of gases is a theory of probability. So are thermodynamics and astronomy. And so is geodesy, involving measurement of the earth's surface for the purpose of figuring the curvature or other characteristics of the earth. It makes use of "least squares."

And I had very good teachers in least squares. That was, by the way, one of the courses I had by correspondence when I was in Colorado.

When I saw some of these things in statistical papers, I recognized them. But in my then juvenile mind, I couldn't believe that anyone could be so naive as not to know the history of the subject. If I just open my old geodesy text at random, you will see things that are familiar, for instance, a result concerning the distribution of the sample variance rediscovered by Gossett in 1908. This text, written in German by Czuber in 1891, was used on the European continent before 1900.

I studied all of these books. When problems came up, I just found myself able to work on them. To help people.

When people had problems with experimental data, I just worked on them and found myself able to make a contribution, of thought anyway. I don't know if I improved the results any. I just found myself doing these things, knowing a little bit about it from the books that I had studied, even though a lot of it had not yet been published in the statistical literature. And I suppose that's the way I got eased into it.

If all this sounds a bit serendipitous, life with Dr. Deming could often be that way. He was able to be at the right place at the right time. He was available, and he was unquestionably the best man for the job. In the ideal Deming world of the early and mid-1980s, each person knew what his or her job was and was no longer hampered by poor tools or inherited defects. Every worker took pride in his or her work and was trained and, when necessary, retrained. Foremen no longer had the responsibility of speeding up production at any cost. The new job was to do whatever was required to help the workers satisfy customers by turning out an excellent product. The company bought not from the lowest bidder, but rather from suppliers who had shown evidence that their processes produced high quality. Quality was, in fact, the overriding management principle.

From the late 1980s until the end of 1993, Dr. Deming gave the world an example of never-ending improvement in his own thinking and his continuous learning. In *The New Economics,* a seminal work published a few months before his death, Dr. Deming summed up his vision of the world now guided by the ability of the human mind to understand the benefits of cooperation and the desirability of what he defined as optimization.

The view of production as a system that he took to Japan now becomes a systematic way to look at the entire organization over time. By optimization, he now means the orchestration of all the interacting components of that system so that everyone, in the long run, will be a winner. Rather than being the guiding principle, quality, in this new view, is simply the necessary outcome of the optimization process.

Thanks for sharing this new vision of the world with us, Dr. Deming. We learned a great deal about the keys to excellence as we studied and grew under you. And you have made our world a better place for having passed this way.

The statistician's job is to find sources of improvement and sources of trouble. This is done with the aid of the theory of probability, the characteristic that distinguishes statistical work from that of other professions. Fighting fires and solving problems downstream is important, but relatively insignificant compared with the contributions that management must make. Examination of the sources of improvement has brought forth the 14 points and an awareness to eradicate the deadly diseases and obstacles that infest Western industry.

W. Edwards Deming
From the Foreword to
The Keys to Excellence

Part Two

THE STEPPING STONES

Chapter Six

CREATE CONSTANCY OF PURPOSE

"Create constancy of purpose toward improvement of product and service, with the aim to become competitive and stay in business, and to provide jobs."

Dr. Deming often emphasized that management is faced with two problems: the problems of today and the problems of tomorrow. He taught that the problems of today must be handled efficiently in order to effectively handle the problems of tomorrow, what he called the future problems. The problem lies in the fact that we spend too much time working on the problems of today, to the exclusion of planning for the problems of tomorrow. Both must be considered and balanced if constancy of purpose is to be achieved and the enterprise is to survive.

This also means the acceptance of obligations and respect for the individual by honoring one's commitments. Although Dr. Deming considered the mission statement as evidence of the beginning of a plan, he often challenged companies with such statements as, "How often do your workers and managers consult it? Is it on the shelf collecting dust?" He would then lay out his "four-legged stool" approach for achieving this constancy of purpose: innovation, research and education, continuous improvement in all that we do, and an effective maintenance plan for furniture, fixtures, and equipment.

In his book entitled *Deming's Road to Continual Improvement,* Dr. William Scherkenbach makes the point that Dr. Deming's philosophy calls for a unique balance between two opposing forces: constancy of purpose and continual improvement—between reduction of waste and addition of value. Scherkenbach goes on to make a crucial point: "But the reduction of waste does not insure value. They are not reciprocals." He further says that Deming's "views on optimization span both space and time," and he points out the need for balance between the individual activities and the team, as well as the balance between short-term and long-term results, for both inputs are important. He likens it to Confucian teaching, in which there is a balance between knowledge and action as well as a balance between science and philosophy. It is in this balance that we find the meaning and value of constancy of purpose.

Mary Walton, in her book entitled *The Deming Management Method,* quotes a conversation with Dr. Deming, the futurist: "People are concerned about the future, and the future is ninety days at the most, or non-existent. There may not be any future. That is what occupies people's minds. That is not the way to stay in business. Not the way to get ahead. You have to spend time on the future." One of the things that made Dr. Deming unique was his innate ability to spend time on the future without it chipping away at and shortchanging the present.

Our next author, Dr. Ernest (Ernie) Kurnow, gives us a firsthand view of this remarkable man's devotion and his constancy of purpose.

"Hello Ernie," and My Day Was Made

by Ernest Kurnow

I first met Dr. Deming almost half a century ago when, as a young instructor in economics (at that time) in our undergraduate school, I was asked to teach a course in statistics at our graduate school. During our first meeting, Dr. Deming asked me to enlighten him about some aspect of economic theory. I cannot recall the precise question. However, I clearly remember his pointing out to me that although he thought my response adequate, it would have been much better if I had indicated the implicit assumptions that I made in my analysis. This encounter served as a lesson to me to beware whenever Dr. Deming feigned ignorance about a certain question. The odds were that he knew much more about the subject than he claimed. Therefore, think carefully before you answer—a habit that has stayed with me for inquiries irrespective of their source.

I never took a formal course with Ed Deming, but if I had to list the best teachers I have ever had, his name would be among them. Ed had a few subtle methods of instruction. Soon after I met him, I received a thirty-page manuscript accompanied by a memo from him. He asked whether the paper was worth publishing, whether I knew of additional references, and whether I had seen Plan II in any paper or book, because he had not. Also, he would be very grateful for any comments by December 1 (about six weeks away). As a young instructor, I was extremely flattered, but after a moment's reflection, I realized that I had been asked to comment on Plan II and I had never even heard of Plan I. I had been given six weeks to research the subject and draft an intelligent response. Now imagine six such memos a year over a period of years and you become well grounded in statistics. This is what I came to know and understand as his point number one—to achieve constancy of purpose.

Ernest Kurnow is a Fellow of the American Statistical Association. He is Professor Emeritus of Statistics and Past Chairman of the Statistics and Operations Research Department at New York University. He was also one of Dr. Deming's first graduate students.

Ed Deming read widely and attended many conferences. When he did not agree with something he heard or read, it was not unusual for him to reply with a note expressing his point of view. I was a regular recipient of copies of these memoranda. He must have felt that as an instructor in economics, I needed a lot of teaching. In these unique ways he taught me, and much of what I teach today I learned from Ed Deming. He also excelled as a morale builder. He would study the programs of the American Statistical Association meetings very carefully. Whenever one of my colleagues or I were on the program, Ed Deming was always in the audience, smiling at us and lending his support and encouragement.

About thirty-five years ago, when my colleagues Gerry Glasser and Fred Ottman and I were writing a basic statistics text, he volunteered to read the manuscript and spent many hours reviewing and analyzing it. His suggestions were invariably valuable and we hoped that we incorporated them adequately. Ed not only helped us with statistics but also reviewed our grammar. I still wince when I see a split infinitive or a noun being used as an adjective (e.g., *sampling* theory or *operations* research). Where Ed found the time with his heavy workload to read the text must remain a mystery. No one in my experience could organize his or her time as efficiently as Ed Deming. He lived the principle of constancy of purpose.

Dr. Deming was a very thorough researcher. He was very careful to acknowledge any ideas that he had received from others. I often came upon interesting problems during an assignment and would discuss them with him. Very often our discussion would find its way into one of his papers or books, and my contribution was acknowledged. All his colleagues and friends were extended the same courtesy. I often think that many of us may well be remembered because we appeared in footnotes of Dr. Deming's books. The following incident illustrates the excessive care Ed took in giving credit to the ideas of others. In 1983, I planned to use his book *Quality, Productivity and Competitive Position* as a text for an executive training program in Greece. Although I had read the text, I did not read the preface until I found time on my hands during the flight to Greece. There, to my surprise, I found the following:

> It was Dr. Ernest Kurnow of the Graduate School of Business Administration who suggested to me years ago that a non-mathematical course teaching the 14 points, illustrating use of tech-

niques with no pretense to teach the techniques themselves, could make an impact on students of management. His idea was sound.

"Years ago" was twenty years ago. I had casually made the remark to Ed one day on our way to class. I had long forgotten about the incident and was reminded of it only after reading the preface to his book. Ed always remembered and constantly and purposefully gave credit where credit was due.

He expected the same respect for sources from all researchers. Many of the copies of memos that I received would point out to writers similar work by earlier scholars that they had overlooked—often articles written before the writer was born. Research, to be respected, had to be thorough. As chairman of the Statistics and Operations Department and subsequently chairman of the Ph.D. program at the Graduate Business School—a span of 22 years—I was in a position to witness the deep interest and affection that Dr. Deming had for his students. Somehow, he always found time to consult with students outside of class, no matter how tight his schedule was. On the rare occasions when he could not make a class, he would make sure that his class was taken over by a well-known statistician or a well-briefed former outstanding student or colleague. I still marvel at the fact that he always asked me, as chairman, for permission to miss his class and informed me of the provision he had made to have his class covered.

For a period of at least ten years, Dr. Deming would meet with students every Friday for several weeks prior to written and oral examinations (held twice a year) to assist them in their preparation. It did not matter where in the world he had to come in from; these Friday meetings were a sacred obligation and you could count on it—he was sure to be there.

As I mentioned earlier, Ed was more knowledgeable in many areas than he would have people believe. During Ph.D. oral examinations, on many occasions I witnessed this knowledge come to the fore in defending a student's reply to a question in another discipline. He would engage in debate with some of our outstanding scholars in economics, finance, marketing, and management to defend a student's reply. They, as I was earlier, were impressed with the extent of his knowledge.

During the first Ph.D. oral exam that I attended, I noticed that a certain professor of management had unnerved a student by asking him esoteric questions about footnotes in texts. The professor, however, was using crib notes which contained the answers to his queries. I did not believe that it was fair for a professor to ask a student to remember a footnote when the professor himself did not know the answer from memory. In those days, a professor without tenure did not speak up, so I informed Ed Deming of what was happening during an intermission period in the examination. He took the management professor to task, and the examination was called off and rescheduled. These are but a few of the many instances that demonstrate Dr. Deming's constancy of purpose and concern for students.

More recently, I would wait after my class on Monday morning for Ed to arrive to meet his afternoon class. Ed would come through the door and greet me with his beaming smile and a glint in his eyes. I would get his firm handshake and warm, deep-throated "Hello, Ernie" and my day was made.

Now Ed is gone and the world has lost one of its great men. My colleagues and I have lost a very dear friend whom we will long remember.

Is every job in a job shop done better than the one before? Is there continual improvement in methods to understand better each new customer's needs? Is there continual improvement of materials, of selection of new employees, of the skills of people at work on the job—and of repeated operations?

W. Edwards Deming
Out of the Crisis

Chapter Seven

END PRICE TAG AWARDS

"End the practice of awarding business on price tag alone and reduce the number of suppliers."

Awarding business based on price alone is one of the most prevalent practices in America today, and Dr. Deming labored unceasingly to eliminate the practice wherever he found it. He often pointed out that five or six suppliers for the same item could multiply problems exponentially. "Another way to put it, defects beget defects...There is too much variation lot to lot and within lots...There is no better way to put it: good quality begets good quality."

Dr. Deming often said that price has little meaning without a true measure of the quality being purchased. He taught that buyers will serve their companies best by developing a single-source relationship with a particular vendor, no matter how controversial that idea may seem to be. In America, we are used to spreading the risk and covering our bets with multiple outlets. Dr. Deming often put it on a practical level: "To work with a single supplier on development of an item demands so much talent and manpower that it is unthinkable that one could go through the development with two suppliers, let alone five or more." Practical wisdom from a practical man.

Dr. Deming laid out a new role for purchasing managers: to change the focus from lowest initial cost of material purchased to lowest overall total cost. He also emphasized that specifications of incoming materials do not tell the whole story about performance.

There are production problems to be dealt with as well. Components and materials may each be excellent on a standalone basis, he pointed out, and yet not work together in production or in the finished product. It was for this reason that he called zero defects a "highway down the tubes."

He often referred to the Japanese transformation with which he was involved as a revolution in thinking, a revolution against fear, and a revolution in supplier quality management. How was it that Dr. Deming went to Japan in the first place? Perhaps we can learn the answer from Homer Sarasohn, who was in Japan before Dr. Deming and who many influential Japanese business leaders believe to be one of the unsung heroes of the Japanese resurrection. Like Dr. Deming, he was not discovered in the United States until his later years and, except to those students of the quality movement who know better, remains hidden in the shadows.

For three or four long years, Sarasohn worked with the military (SCAP) under General MacArthur to effect the transformation of the electronics industry. The results were spectacular. Yields increased from ten percent to ninety percent and the standard of living rose dramatically. Yet he received no fame or fortune for his efforts. About six months ago, he and his wife came to my home for dinner and I posed the following question: "Homer, doesn't it ever bother you that Deming and Juran got most of the glory and, except for the old-timers in the Japanese business community, very few people have heard of you and your work?" He thought about it for a minute or two, and his answer speaks volumes about the man.

He told me that in order to be effective in Japan, he had to learn the language, for it seems that the Japanese were extremely polite to him and said yes even when they meant no. He learned the language in order to communicate directly, instead of through an interpreter, so that he would know what the leaders were saying and thinking. In doing so, he was able to introduce new words and concepts into the Japanese language. He summed it up by saying to me, "Do you think I care who gets credit for the successes and the glory when I've been able to add words and ideas and concepts to an eighteen-century-old culture and influence their thinking?"

I know you will enjoy his story and insights, for he is a genius of a man whom I have come to admire and respect so much. He was "present at the creation"—at the recreation of the Japanese economy—and is a firsthand witness to the power of the transformation.

The Return of the Herald Angel

By Homer M. Sarasohn

William Edwards Deming's death should not be mourned. Rather, his life and teachings should be celebrated. Although frequently cantankerous, outspoken, acerbic, and opinionated, at heart he was always a sincerely motivated idealist.

It is only within the past fifteen years or so that Dr. Deming has been acknowledged in the United States and in Europe as considerably more than a "consultant in statistical studies." That designation oversimplifies and minimizes the major contribution of his life. What he actually accomplished, using the quality process as a tool, was a breakthrough to the conscience of hidebound, status-quo-oriented industrial leaders. He looked beyond mere statistics and saw the process of controlling quality of performance as the mechanism by which the functions of management could be continuously improved. It is from this perspective that modern concepts and principles of total quality management have evolved.

Dr. Deming's education and training in physics and mathematics provided him with the insight, from a scientific viewpoint, to comprehend the interconnectability and interdependence of disparate but related elements. From this he recognized the essential nature of a total system. A manufacturing operation, for example, was not a succession of individual events, as was the common conception in the early 1900s. It properly must be regarded and treated as a continuous interrrelated stream of activities. It was his exposure to the genius of Dr. Walter A. Shewhart that brought this awareness to a focus.

He came to know Dr. Shewhart when he was engaged as a part-time employee during the summer of 1925 in Chicago at the Hawthorne plant of the Western Electric Company. Shewhart, a physicist, had

---

Homer M. Sarasohn is an electronics engineer who went to Japan in 1945 following World War II to help with the transformation of the Japanese economy. After consulting with Booz, Allen and Hamilton, he became Corporate Director of Engineering for IBM. He is currently a hearing officer in the Court of Arizona and a member of the Arizona Quality Council.

been given the task of solving a major manufacturing problem. Western Electric's inspection engineering department had been working for quite some time, without success, to bring the products being produced at the plant into an assured state of conformity with specifications and uniformity. Nevertheless, there continued to be wide variation in acceptability from one batch to another. In fact, the problem was becoming even worse as the inspection team fumbled for a solution.

The situation was particularly perplexing. In the early 1900s, the American Telephone and Telegraph Company (AT&T) was expanding its communication system across the country. It was becoming technically more complex as more and more apparatus of all kinds was added. Nevertheless, AT&T's underlying commitment was to ensure the reliability of the total system, and every part of it, in order to earn the confidence of its subscribers (customers), control costs, and reduce materials waste. The Hawthorne plant was the major supplier of components (switch gear, telephones, etc.) to the AT&T network. If it could not produce products of dependable uniformity, the parent company would be unable to keep the trust of its customers.

As Shewhart studied Hawthorne's manufacturing operations, he became aware of two distinct types of causes of system instability. One was various random variations that occurred within a process itself as it was designed, installed, and operated. He named these *common* causes. The other were *special* causes that could be identified as separable from the process as designed. However, they affected the process in ways that made its output become unpredictable. These special causes gave rise to wider and more aperiodic output variations than would be expected to be the result of common causes.

On May 16, 1924, Shewhart reported the results of his investigation to the head of the inspection engineering department. With his note, he enclosed "a form of report designed to indicate whether or not the observed variations in the per cent of defective apparatus of a given type are significant; that is, to indicate whether or not the product is satisfactory." This enclosure was the first of Shewhart's soon-to-become-famous quality control charts.

What attracted Dr. Deming to Shewhart's conclusions was, first of all, his mathematically valid method for determining the stability of any manufacturing process. It thus became possible to identify the assignable causes that influenced the process output to wander into unpredictability. With such information in hand, it became practical and useful to make a determination as to what corrective action was needed to recover system stability, as well as the economic value of such action or any alternative action. Thus, the process could be improved, costs could be controlled, manufacturing productivity could be increased, and the value of the output to the enterprise and to its customers could be enhanced.

However, unless the distinction between common (unassignable) causes and special causes and the separate effects of each were recognized, in addition to targeting the corrective action accordingly, then the process control problems could only become worse. Failure to recognize the separability of such causes is what happened at the Hawthorne plant. The inspection department had been trying to correct the difficulties it perceived without really understanding what the nature of the problems was. Its efforts were thus to no avail, and the problems got worse.

Dr. Deming recognized some other realities. Using Shewhart's control charts and his three sigma standard deviation, special causes could be identified and brought under control by process supervisors, whether it be the need for better operator training, adjustment of the process environment, attention to the inflow from suppliers, or other assignable factors. The common causes that lie within the system, however, would remain even after such special causes were eliminated. These causes could probably be controlled and reduced and the system thus further improved (at some cost in terms of time, effort, and money), but it would take the awareness of management to see the need for continuous improvement of the system, and it would take the reaffirmation of leaders to the commitment of satisfaction of the customers to put the necessary changes into effect. This is where Dr. Deming saw the ultimate goal that stood beyond statistics: *The attainment of the quality objectives of an enterprise is the responsibility of its management team.*

He went beyond Shewhart in another matter. Shewhart continued his research in statistical process control at AT&T's Bell Laboratories. His interest was centered primarily on the manufacturing environment. His book *Economic Control of Quality of Manufactured Product* was published in 1931. Dr. Deming saw that Shewhart's methodology could be successfully employed in any systematic process. That was what interested him. After his stint at Hawthorne, he returned to his graduate studies at Yale University, got his doctorate in mathematical physics, and then was employed by the U.S. Department of Agriculture. Later on, he went to work at the National Bureau of the Census as a statistician specializing in the development of sampling techniques. One of his major contributions was the adaptation of statistical process control methods to such tasks as data collection and data processing in preparation for the 1940 national census. Reportedly, he was able to gain operations improvements and error reductions of up to six times over previous records. The payoff was a substantial increase in clerical productivity and attendant dollar savings in addition to an earlier release of the population census results.

In 1942, Dr. Deming was engaged by the U.S. Army as a consultant to set up courses in statistical process control methods to be taught to engineers and others involved in the production of ordinance material. Here again, he and those associated with him in this effort were successful. Productivity gains were substantial, and reduction of manufacturing rejects and waste was notable. However, these benefits did not carry over beyond the end of World War II to production for the civilian market. Manufacturers quickly shifted their facilities to take advantage of the pent-up consumer demand, which took priority over any other consideration. For them, quality was not an issue.

In her book entitled *The Keys to Excellence,* Nancy R. Mann quotes Dr. Deming as follows:

> The [wartime] courses were well-received by engineers, but management paid no attention to them. Management did not understand that they had to get behind improvement of quality and carry out their obligations from the top down. Any instabilities can help to point out specific times or locations of local problems. Once these local problems are removed, there is a process that will continue until someone changes it. Changing the process is management's responsibility. And we failed to teach them that.

When the war in the Pacific came to an end toward the end of 1945, with the Japanese surrender to General Douglas MacArthur on the deck of the battleship USS Missouri, the American government had not yet decided what course it would follow with that war-devastated nation. Its economy was at a standstill. Its productive capability was demolished. Its work force and sources of raw material were no longer viable. In some areas, people were actually starving—there was not enough food to go around. Rather than letting Japan become dependent upon American taxpayers, President Truman decided that Japan, under MacArthur's command, would be re-established and rehabilitated as a fully functioning democratic economic entity.

One of the first priority actions taken as the post-war U.S. occupation of Japan got under way was the rebuilding of the country's communication systems and the manufacturing industry that supported and supplied those systems. Total responsibility for this effort was placed in the Civil Communications Section (CCS) of MacArthur's General Headquarters (GHQ) staff. As a member of the CCS Industry Branch, I had charge of restarting the communications equipment manufacturing industry quite literally from the ground up. That re-establishment process—relocating of factories, refurbishing of machinery, reorganizing of production, retraining of workers, restaffing of operations, and all the other things that had to be done to recreate an industry—continued for five years. By 1949, reliable production was in operation and the requirements for quality control were established in the industry for the first time in Japanese history. To cement those gains and changes from Japan's feudal past, however, two more things had to be done.

First, an Electrical Test Laboratory was set up within the Japanese government to enforce the quality standards that we in CCS had imposed upon the manufacturers. Next, we set up a university-level mini MBA course, the CCS Seminar, for senior executives of the industry, who, along with certain university professors and ministry officials, were required to attend. Upon "graduation" from the seminar, these executives were required to repeat the course for their subordinates. The course continued, sponsored by the Japan Management Association, long after the occupation ended. The textbook that Charles Protzman and I wrote for the course is still available there and was republished recently in April 1993.

Our CCS plan called for the seminar to be followed by further intensive training in statistical quality control and total quality management methods. That plan was interrupted, however, by the advent of the Korean War in the summer of 1950. Since we in GHQ suddenly became otherwise occupied with that matter, I looked for someone else to take up the burden of carrying on with what we had started in the industry. Japanese managers and workers were already responding to the new style of management with honest dedication. That momentum could not be lost. The person selected to take over, therefore, had to be a Shewhart disciple, if not Shewhart himself. We contacted Dr. Shewhart, who was ill at the time and could not come. However, Dr. Deming agreed to come and the die was cast.

On his arrival in Tokyo, Dr. Deming was met by a group of engineers and scientists, some from a recently formed Union of Japanese Scientists and Engineers, others from an older Group for Statistical Research, and still others from the Japan Efficiency Association. To some degree, these people were generally knowledgeable about Taylor's Scientific Management principles. From materials we in CCS had given them, they were familiar with Shewhart's work. Now, they saw the chance with Deming to get to the theories and fundamentals of statistical process control that had "won the war for America."

Although courteous, Dr. Deming did not see his role as teaching the mathematics of statistics. He already had the experience in the United States of dealing with technical specialists and missing the chance to reach and influence corporate decision makers. At the end of the war, he had gone to New York University as a professor in the School of Business Administration. He had also set himself up as a management consultant. As to the contributions he had made to wartime productivity and other gains elsewhere in government, he was practically an unknown in his own country. In fact, his associates at the time dubbed the years from 1946 to 1948 a "blue funk period." Not only were the clients not knocking his door down, they were not even tapping loudly. One colleague had remarked to me that he seemed almost depressed by the failure of American hidebound management to adopt a constancy of purpose. How quickly they returned to the old practices and ways.

Nevertheless, he had proof to his satisfaction that the tangible benefits of total quality were obtainable. The key that opened the gate to the possibilities was top management commitment. Within the organization, from the top down, the people had to be convinced of the sincerity of that commitment. That was the message he wanted to bring to Japan, and he sought out the top executives of the leading Japanese companies as his audience. At first, they listened to him with customary politeness. Then, they began to understand the challenges and opportunities that might be theirs if they adopted his methods. He laid out a plan of action that they could envision. It carried, if followed conscientiously, the potential for tangible attractive rewards. In his initial lectures, and in the subsequent visits he made to Japan, Dr. Deming captured the attention of his listeners. They became converts to his philosophy with an enthusiasm that even he did not expect. Along with other factors (political, economic, technological, etc.) that aided its advance, in a relatively short time Japan emerged upon the world scene as an industrial power to be taken seriously.

Belatedly, American industrial executives, becoming aware of Japan's progress, began to realize that their companies might also profit from a change from their status quo mentality. Then the question was asked: "If Japan Can...Why Can't We?" The documentary by that name, broadcast on the NBC television network in 1980, brought Dr. Deming to national attention.

Before too long, he was accorded the recognition he was entitled to have received quite some years before. To the day of his death, he was in demand as a teacher and lecturer. People wanted to hear what he had to say. They listened and they learned. He responded to their demand, even in his later years when he was physically incapacitated. He had a message to give: "Be not afraid of change; seek always to improve. And most importantly, as Churchill once said—never, ever ever, ever ever give up!"

> *Wherever I taught, I told the people nothing will happen if you don't keep working together, and you've learned only a little, so you must keep on working and meeting together.*
>
> W. Edwards Deming
> *The Keys to Excellence*

Chapter Eight

ELIMINATE NUMERICAL QUOTAS AND WORK STANDARDS

"Eliminate work standards (quotas) on the factory floor. Substitute leadership. Eliminate management by objectives. Eliminate management by numbers and numerical goals. Substitute leadership."

Dr. Deming believed that standards, piecework, and quotas are crutches upon which poor supervision rests. They impede quality more than any other single working condition. For the most part, they guarantee inefficiency and high cost. Because standards often contain allowances for rework and defective items, it is a surety that management will receive mediocre performance at best.

The problem with standards and quotas, according to Dr. Deming, is twofold. First, he taught that even if the standard is fair, approximately half of the people will fall below and half will be above the average. However, peer pressure usually holds the upper half close to the average rate, while those below cannot possibly attain the average, thus resulting in loss, chaos, poor morale, and turnover. The second problem is that quite often management sets the standard toward the high side in order to create "stretch goals and weed out those individuals who can't make it," thus lowering morale even further. In either case, standards are often self-limiting and workers learn to stop when they have achieved the goals, which results in idle time.

As much as Dr. Deming disliked standards, he liked piecework even less. He said that this system encouraged workers to turn out volume rather than quality. Unless the cost of rework and rejects is factored into the equation using some form of statistical process control, the end result is often lower productivity and poor quality.

In order to combat these conditions, Dr. Deming taught that it is the job of the supervisor to help workers perform better quality work and to increase productivity on a continuing basis. To accomplish this, they must have a good understanding of the work, coupled with an appreciation of the capability of the process using statistical process control techniques. Thus, he felt that every organization needed the in-house services of either a part-time or a full-time statistician.

Dr. Howard Gitlow, who has been a follower of Dr. Deming and his philosophy for twenty-five years or so, offers us a glimpse of the make-up of the statistician and the multiple roles that this individual must play, as personified by his mentor, Dr. Deming. I first met Howard in the mid-1980s when Florida Power and Light was involved in the creation of the Quality Institute at the University of Miami, where I was an adjunct lecturer in the evenings and on Saturdays. We became friends over the years, and his ability to articulate the complexities of statistics and total quality management in simple and understandable terms never ceases to amaze me. His insights and comparative analysis during and after the Florida Power and Light Deming Prize challenge were a tremendous source of encouragement. His insights into the person and work of Dr. Deming make the next tribute come alive with poignancy and meaning.

Thanks for the *Michi*

by Howard S. Gitlow

In October of 1969, I was president of the Statistics Club at New York University. In that capacity, I was called to Dr. Deming's office so that he could ask me a question. I had never met him before, but knew that he was a great man. I put on my best suit and combed my hair into a ponytail. Upon my arrival, Dr. Deming looked at me and said, "Why don't you get a haircut?" I was completely disconcerted. I answered his question and immediately left his office.

Eleven years passed, during which I had no contact with Dr. Deming. In 1980, after NBC aired "If Japan Can...Why Can't We?," the dean of the School of Business Administration at the University of Miami called Dr. Deming and asked him to come to Miami to talk about quality. The dean also mentioned that one of Dr. Deming's former students now taught at the university. Dr. Deming asked who, and the dean replied, "Howard Gitlow." Dr. Deming said, "Tell me, do you pay him enough so he can get a haircut?" What a memory! Once again, I was to meet Dr. Deming.

In preparing for Dr. Deming's visit to the University of Miami, I asked him if I could pick him up at the airport. He told me I could pick him up, but not at the airport. I should meet him at the train station. I asked him why not the airport, to which he replied, "You cannot travel like a gentleman on an aeroplane." I thought, "Okay, here I go again." I was ready and waiting for him at the train station, only to discover that the train had arrived early and Dr. Deming was nowhere in sight. I panicked and started running around the station, calling his name. In frustration, I drove to the hotel. There he was, cool as a cucumber, waiting for me. He was very nice, but once again, he got me frazzled.

Dr. Deming surprised me by asking if I would sit on the stage with

---

Dr. Howard S. Gitlow is Executive Director of the Institute for the Study of Quality in the School of Business Administration as well as Professor of Management Science at the University of Miami. He holds a Ph.D. in statistics from New York University and has written extensively in the field of total quality and statistics.

him while he delivered his four-day seminar. I agreed and was flattered that such a famous man was paying attention to me. I think I listened a little harder than I would have otherwise. I heard his message; it was titillating. Right then and there, I made a career decision: I would give up research in traditional multivariate statistics and study Dr. Deming's ideas. They were unusual, but I thought maybe I could make a mark for myself in his brave new world. As a consequence of our second meeting, I began one of the most significant journeys of my life, a journey of personal and professional transformation.

Over the years, I have had the privilege of traveling with Dr. Deming. He was always concerned and treated me with respect. His ability to focus on issues and remember details was remarkable. His energy and stamina never ceased to amaze me. He was an anchor for all who wanted to learn and improve. He was fearless. I recall being at one of America's largest firms with Dr. Deming on his eighty-first or eighty-second birthday. The top management of the company had arranged a birthday party, complete with a huge cake, in his honor. After cake and coffee, Dr. Deming addressed the top management. His opening lines almost knocked me out of my seat. He said, "Do you know what is wrong with your company?" The room was silent—you could hear a pin drop. Dr. Deming pointed to the president and chief executive officer and said, "Him. He is what's wrong with your company." And so my lessons as a consultant began.

I must confess that, to my way of thinking, he was not a role model for healthy eating. I recall breakfasts of toast with bacon and salt, plus many cups of coffee, and dinners of onion soup, napoleons, and key lime pie, with two piña coladas with extra jiggers of rum. Even after all that, he would function like the master, always asking tough questions, trying to improve me. Sometimes it wasn't easy, but it was always interesting, thought-provoking, and mind-expanding.

Dr. Deming taught me theory. He taught me how to ask questions and learn. He taught me that theory is practical and can help people improve their daily lives. Dr. Deming did not believe in spoon-feeding his students. It was the Socratic method all the way. I remember calling him in 1981 or 1982 with a question which would clarify six months of personal study. Instead of answering my question, he said, "You're barking up the wrong tree." It took me six more months to figure out why I was barking up the wrong tree and which tree I should be barking up. Once I figured it out, I realized that I had learned more by his method than I could have by any other means. It wasn't easy, but I learned.

It was because of Dr. Deming that I began to study *michi,* a Japanese term used to describe the footsteps a person must follow to pursue cosmic oneness and peace with the universe. I study the *michi* of the statistician, which is a path that provides understanding through the application of statistical theory to the processes of life.

Michi must be learned from a master. Dr. Deming was my master. He was my guide on the path to transformation, a discontinuous path with peaks and valleys. He guided me through the valleys and helped me over the peaks. I began to transform and perceive new meaning in my life and events. Dr. Deming taught me how to create joy in learning and life. He taught me the System of Profound Knowledge.

Dr. Deming taught the principles of statistics. I applied those principles to my life and improved myself. An example follows. When my daughter was eight weeks old, my wife and I joined a play group called "Having Babies After Thirty." It was a wonderful support group for new parents. After about four months, the parents began to seat the children around a table and give them apple juice. The six-month-old children would flop around in their seats. Sooner or later, each child would knock his or her apple juice off the table and onto the floor. The parents fell into two groups in terms of their response to the spills. The first group said nothing, quietly cleaned up the spill, and gave the child more juice. The second group yelled at the child ("Be careful—don't spill the juice!"), cleaned up the spill, and gave the child more juice. This went on for approximately four years. Although I never kept charts, at some point I realized that the children of the first type of parent spilled less juice (on average) than the children of the second type of parent. I thought about this a great deal, until I realized that I was confronted with a fundamental principle of statistics which I had learned from Dr. Deming.

The first type of parent was subconsciously treating the spills as a system of common cause events; that is, they were treating the apple juice spills as a process—the apple juice spilling process. Over time, these children seemed to have fewer and fewer apple juice spills. I believe that this was a result of the parents subconsciously recognizing that spilling apple juice was due to common causes of variation which were beyond their control.

The second type of parent repeatedly treated each spill as a special cause of variation from the desired state of no spills (zero defects). They were overreacting to the common cause events. Over time, these children seemed to spill more and more apple juice. I postulated that this was due to the parents pushing their children beyond their capabilities.

Dr. Deming taught me about the two type of variation in a process, chaotic and stable processes, and the management of each type of process. This is part of the *michi* of the statistician. I was able to apply this *michi* to parenting. Dr. Deming taught me not to be an overreactive parent who demands behavior beyond the capability of my child. Rather, he taught me to be a parent who understands when my child's behavior is stable and what to do in that situation.

People are not born knowing *michi*. It requires years of study and tutelage under a master. Dr. Deming was my master. I thank him for teaching me the theory of statistics so that I could pursue the *michi* of the statistician. Dr. Deming, thanks for the *michi*.

The problem is to improve the system and find out who is having trouble. Isn't it clear? Numerical quotas—so many per day; a plant manager—so many per day. If he fails to meet it, he fails. No regard for what is a day's work. No possibility to improve. Do you think a plant manager will report 7,000 when the quota is 5,000? That he will report 5,000 when the quota is 5,000? No! Put them under the counter. We may need them for a rainy day. It may rain tomorrow.

W. Edwards Deming
The Deming Management Method

Chapter Nine

CEASE DEPENDENCE UPON MASS INSPECTION

"Cease dependence upon mass inspection to achieve quality. Eliminate the need for inspection on a mass basis by building quality into the product in the first place."

Dr. Deming said that quality comes from improvement of the process, not from inspection of the completed product, because inspection is often done too late and is costly and ineffective. By performing routine inspection, a company plans for rejects and defects. The basic problem is usually that the process is out of statistical control or the specifications are not workable.

He emphasized that the underlying problem is often not the quality of the product, but rather the quality of the process. The solution includes providing workers with information and training so that they can inspect their own work on a routine basis. Dr. Deming also emphasized the need to create the right organizational climate that will allow workers to use their judgment based upon statistics in order to take corrective action. In order for each worker to be responsible for his or her own quality, management must first believe that employees want to do their best. Furthermore, in order to do better, they must be given the tools, methods, and opportunities with which to do so.

He acknowledged, however, that there are exceptions and circumstances in which mistakes will inevitably occur. He used as an example the manufacture and fabrication of an integrated circuit, which is complicated and must be inspected to weed out the good ones from the bad. This example emphasized that it is important to carry out inspection at the right point for minimum total cost.

His theory, as laid out in Chapter 15 of *Out of the Crisis,* was to plan for the minimum average total cost for test of incoming materials and for final product. He taught the "all-or-none" rule which governs incoming inspection. He taught about the value of 100 percent inspection to accumulate information as quickly as possible. This is especially necessary when in a state of chaos. His four rules to govern the inspection process are as follow:

- **Rule 1:** Inspection does not improve quality or quality products, because it already too late.
- **Rule 2:** Mass inspection is unreliable, costly and ineffective and should not be used, except in a state of chaos.
- **Rule 3:** Inspectors rarely agree with each other and are easily bored, and their instruments are a pain in the neck, because they require continual maintenance.
- **Rule 4:** Sampling using control charts allows for comparison and a professional job.

The next contributor is Dr. Gerald Glasser, who was a student of Dr. Deming's in the 1950s and continued under his tutelage thereafter. He saw Deming as the innovator, always introducing new inspection and sampling and statistical techniques, as well as modern behavioral methods, into the classroom, long before the techniques were popular. He was able to see firsthand how the application of the sampling method could transform inspection practices not only in the workplace, but in the classroom as well. At the suggestion of our mutual friend Howard Gitlow, I approached Dr. Glasser about writing a tribute for this book. His response both surprised and encouraged me: "Enclosed is a tribute to Dr. Deming that I gave at his 75th birthday celebration in 1976. He had a copy framed and hanging on the wall of his New York apartment for many years. Let me know if it is suitable for your book." He concluded with the following encouragement: "The project is very nice. Thank you for asking me to participate. Best wishes, Gerry."

My own memories of New York City returned after reading his letter. In 1976, I was an occasional part-time lecturer at New York University, where I filled in for my boss at the time, Bob Wetzler. I recalled a few delightful moments spent in Dr. Deming's New York apartment with my teacher and mentor at St. John's University, Dr. Jean Namais. How proud Dr. Deming must have been of Dr. Glasser's tribute and the words it conveyed, which are just as pertinent today as when they were first written almost twenty years ago.

Following Dr. Glasser's tribute is a remembrance by Hana Tomasek, one of the first people to bring the new philosophy to Eastern Europe in early 1990. She is a pioneer who took large risks to bring the quality gospel back home to her native country but who never forgot to send Dr. Deming flowers on his birthday.

A Tribute to W. Edwards Deming

by Gerald Glasser

It is a fine honor for me to participate in this tribute to W. Edwards Deming. Dr. Deming has always taught the importance of precision. Hence, I have written out my remarks. This will enable me to be precise and, as a side benefit, it will enable me to be concise.

Ed Deming represents many things to many people. Everyone knows him, or knows of him, as one of the leading statisticians in the world over the past forty years. It is likely, as we meet here, that students in several countries are reviewing books or papers that he has written. It is even more certain that people throughout the United States, Europe, and Asia are at this very minute carrying out instructions as part of statistical studies designed by W. Edwards Deming.

Dr. Gerald Glasser is a Fellow of the American Statistical Association and a principal in the firm Statistical Research, Inc. of Westfield, New Jersey.

In thinking about what I could say, I recognize that it would have been very easy for me to be clinical, to logically and systematically comment on his many accomplishments and contributions. To do this properly, of course, would have taken hours—for his list of credits includes several great books, approximately 150 papers, many significant prizes and awards, an impressive list of clients, and a career as a great teacher. I prefer, however, to assume you are all mostly familiar with those accomplishments and contributions and to talk with you about Ed Deming on a more personal level.

Each of us has had a unique association with Ed Deming, and I can hardly find the words to communicate properly my own association with him, let alone try, as I should, to speak for the group. So let me, *a priori,* admit to inadequacies in how I shall express myself, but nonetheless try to share briefly some glimpses of Ed Deming that I garnered over the past quarter century.

The fact is he is not only a great statistician. He is a great human being, and I relish the fact that I know him as a friend, as well as a colleague. Mostly, of course, I think of him as my mentor, not only because I first came to know him as my professor, but because to this day I turn to him for advice and counsel on the tough problems I meet, technical or philosophic or both. My tutelage continues and will continue, and I am very proud of it.

It was in 1953 that Ernie Kurnow, who also has had a profound influence on my life, talked me out of donning a green eye-shade and becoming an actuary at the Metropolitan Life Insurance Company. "Go downtown to GBA," Ernie said, "and study statistics with Professor Deming." "Why?" I replied with all the infinite wisdom of a twenty-one-year-old, "Don't I know everything already? After all, I had four courses with you." As you all know, Ernie always wins an argument. Fortunately, I went to GBA, to the old Trinity Church Annex building, to study statistics with W. Edwards Deming.

We have a great school in GBA today, but it was great in a different way back in the 1950s. The building was ancient. The classrooms were worn. No elevators. The top floor of classrooms had peaked, thirty-five-foot ceilings and rafters. I took classes in the basement from Professor Deming and Dutka and Frankel—no more than ten or fifteen feet away from the boiler. The faculty consisted of several great men, and students came to study at their feet, as the saying goes. Virtually all of the students agreed on one point: it was the finest education anyone ever had or could have.

Classes with Dr. Deming were interesting and exciting. He stressed theory and showed why it was important by demonstrating how it could be applied in the real world. He introduced visual aids in the classroom before they became the "in" thing. I remember well the bead-box experiments we conducted in his classes to demonstrate the theory of sampling and principles of quality control. Dr. Deming's bead box was famous. Many stories about it circulated among the GBA students—how carefully he handled it, how important the counts were, etc. In my first class, I drew the first sample. Plop. A bead dropped on the floor. My career as a statistician almost ended.

Not to detract from Ed's many virtues as a teacher, I feel special mention should be made of the precision with which he writes text and formulas on a blackboard. Few professors, I am sure, have ever been neater and none has ever written in a straighter line.

I remember clearly one class with him. A student asked a question. A few thought it unnecessary or dumb and giggled unkindly. Dr. Deming handled it as the gentleman and scholar that he is. Before going on to answer the question, he responded: "That's an interesting question." This merely illustrates one of his many virtues. Dr. Deming, busy as he is, has always been courteous, respectful, and responsive to students. His greatness derives not only from the fact that he has important things to say; it derives from the manner in which he says them.

In class one day, he said, "There is only one way to learn statistical theory. You must work with a master." Little could I have then dreamed that in a few years I would have that opportunity. Starting in 1962, he asked me to work with him on several consulting assignments. I was proud to be his assistant, though he never treated me as an assistant. I was his associate. My name appeared with his on papers and reports. I was receiving the finest statistical education possible—and being paid for it too!

What did I learn? Many, many things, including how to design sampling plans and how to make certain those plans were followed; how to deal with clients, how to divide responsibilities, how to deal with lawyers, and how to testify; how to work weekends and evenings, and how to not only enjoy, but to relish, my work; the joy of learning something new ("Every job is different," he would say, "try to learn something new every day."); how to learn from others (Just a few days ago he sent me a paper with a note, "You could help me by being severe on criticism."); how to maintain a high level of professionalism, and how to work so you can look back at what you've done with pride and no regrets; pride in my work and in being a statistician; and last, but not least, how to budget my time. I will recall a day Dr. Deming had out his famous calendar, trying to set a date with me to discuss some work. "Is next Thursday possible, Ed?" I asked. "Well," he responded, "Let's see. I'll be in New York that morning and have to be in White Plains by 8:30. After that, I'm flying to Rochester and Buffalo for meetings at the airport with lawyers. Then it's on to Cleveland and Akron in the afternoon. But let me see what we can fit in."

I have been selfish today in focusing on my own relationship with Dr. Deming. Multiply this by a very large number and you have his contribution to his students—and to society.

W. Edwards Deming is friend, colleague, and mentor to me, as he has been to many. We are not only proud of it, but grateful for it. He has had a profound influence on many careers and on many lives. At this point, I can add only one thing. Ed, thank you. Thank you very much.

Remembrances from the Czech Republic

Hana Tomasek

I met Dr. Deming ten years ago, when I first attended his four-day seminar. He was full of energy at that time, eighty-three years young. I fell in love with him right away. I admired his knowledge and his devotion to the world and, above all, to America, my beloved country. I am still angry with American managers who, around 1950, did not listen to this wise man until it was thirty years too late and

Dr. Hana Tomasek is one of the leaders of the TQM revolution in Eastern Europe and is president of her own consulting organization, INCOS, located in Bloomington, Minnesota. She is also co-founder of the Czech consulting group Inventa.

America was in trouble. Will we ever get out of trouble? I don't know—we do not have Dr. Deming anymore.

I adored Dr. Deming as a man. In private one-on-one conversations or at dinner with friends, he was a gentle, loving person, gracious and courtly in the respectful, old-fashioned way of a true gentleman. I also agreed with him when he screamed at participants at his seminars. When participants would bring up examples of problems within their companies, Dr. Deming sometimes raised his voice, "...was it the fault of a worker? No, it was a fault of a system. And who is responsible for the system? Management..."

I have tears in my eyes whenever I read the letter from a woman who attended one of Dr. Deming's seminar. He included it in his last book:

> A Willing Worker named Ann, after the experiment on the Red Beads came to a close, expressed to me some provocative thoughts. Please put these thoughts into writing, I pleaded with her. Please write them just as you told them to me. She did. Here is her letter.
>
> "When I was a Willing Worker on the Red Beads, I learned more than statistical theory. I knew that the system would not allow me to meet the goal, but I still felt that I could. I wished to. I tried so hard. I felt responsibility: others depended on me. My logic and emotions conflicted, and I was frustrated. Logic said that there was no way to succeed. Emotion said that I could be trying.
>
> "After it was over, I thought about my own work situation. How often are people in a situation that they can not govern, but wish to do their best? And people do their best. And after a while, what happens to their desire? For some, they become turned off, tuned out. Fortunately, there are many that only need the opportunity and methods to contribute with."
>
> W. Edwards Deming
> *The New Economics for Industry, Government, Education**

Dr. Deming taught me so much, but he taught me about systems thinking in particular. I have applied this knowledge to any and every situation, both professional and personal. For example, I blame the system for children failing in school. Both the school system and the system of raising children at home are responsible for what is happening to our children.

* Massachusetts Institute of Technology, Cambridge, 1993, pp. 167–168.

I also learned modesty and the art of persuasion from Dr. Deming. He did not care for awards, and although he must have been frustrated and disappointed when American managers would not listen to him earlier, his love of the United States and the world would not allow him to stop trying to teach.

Several years ago, I was fortunate enough to be Dr. Deming's helper at one of his seminars. I will never forget the experience. It was at the time of the "Velvet Revolution" and the fall of Communism in the former Czechoslovakia, the country in which I was born. He asked me to speak to the seminar participants about the big change there. Dr. Deming was aware that I had started to teach his philosophy in the former Czechoslovakia (now the Czech Republic and Slovakia) in 1990, and I continue to do so, even though sometimes I am almost laughed at. Some managers say, "It will not work here," and I can almost hear Dr. Deming saying, "How do you know? On what data do you base this? Have you tried it?"

Contrary to what some people might think, Dr. Deming had tremendous respect for people, sometimes to the point of risking his own health to accommodate them. While in Minneapolis in 1992, he had invited a few friends, including myself, to dinner. We were waiting for him to arrive when we received word that he was not feeling well and had checked into a local hospital. Shortly thereafter, he walked into the restaurant. He did not want to disappoint us. Even though he was on a special diet, and had to forgo his usual gin and tonic, he spent an hour with us, and we talked about my work in the Czech Republic and Slovakia. Some time earlier, I had received a letter from him saying, "...some day we will go to Prague together." He was ninety-one at the time. In Atlanta, during another dinner with his friends, we sat hidden in a corner, so that he would not be recognized. He wanted to concentrate on his friends. He always treated the waiters and waitresses courteously, and I am sure he was very generous to them.

Even as he was dying, Dr. Deming was thinking of others. It was his request that any memorials be made in the form of blood donations to Sibley Hospital in Washington, D.C. While I was donating blood, I learned that Dr. Deming had told his nurse, "Life is so fragile. Let's not be afraid to say 'I love you.'"

I feel fortunate to have had the privilege of knowing Dr. Deming, not only as a world-famous guru but also as a person. I miss him very much. I will never again send him flowers for his birthday, but I will always have them for him in my heart.

The moral is that testing may cause more trouble than the product itself. Much product is falsely condemned in industry only because of measurement processes that give answers that do not agree with other answers.

W. Edwards Deming
Out of the Crisis

Part Three

THE MILESTONES

Chapter Ten

BREAK DOWN BARRIERS BETWEEN DEPARTMENTS AND ORGANIZATIONS

"Break down barriers between departments and organizations. People in research, design, sales, and production must work as a team, to foresee problems of production and in use that may be encountered with the product or service."

Dr. Deming often told stories about departments within companies that did not get along due to lack of understanding of the other area's problems and limitations. One of the classics, the parable of the shoes, is relayed by Mary Walton in *The Deming Management Method.*

It seems that a certain company had a technical R&D staff that was very creative, so creative that they came up with a breakthrough shoe design that was sure to give them a competitive edge. The staff made ten prototype pairs for the salesmen to show their customers. The customers loved the shoe, so much in fact that over 10,000 orders were placed in one week. Instead of being overjoyed, the situation turned sour quickly. There were so many orders that production could never fill them in a month, much less in a week, as promised by marketing. The salesmen had to go back to their customers and tell them that the orders could not be filled. Hypothetical? Absurd? Not really. According to Dr. Deming, the same scene unfolds every day in every way across corporate America.

Management is often part of the problem when they make spur-of-the-moment design changes or delivery commitments and then leave the responsibility for making it happen to their subordinates. The solution in part is using cross-functional teamwork and task forces, even between organizations, in order to break down barriers and connect the pieces. In fact, whole communities are being connected through quality councils and coalitions to help solve one another's problems.

The Philadelphia Area Council for Excellence (PACE) is one of the earlier community councils created in America, whose purpose is the furtherance of improved quality and productivity as a shared concept and responsibility by each of its members. Their declaration begins with the words, "We shall create a constancy of purpose for improvement of products and services" and continues for nine or ten such statements, ending with, "It is essential to eliminate fear in order to foster a creative environment." Sound familiar?

Mary Ann Gould was one of the original founders of PACE and one of the early quality leaders in the Philadelphia area community. Maureen Glassman is executive director of PACE and has been there from the beginning in 1985. I first met Maureen in 1986, when Florida Power and Light had purchased the Colonial Penn Insurance Company and the then-chairman Marshall MacDonald sent Kent Sterett and myself to Philadelphia to help teach them the Quality Improvement Process (QIP) that we were showcasing. We had decided to go for the Deming Prize, and Marshall wanted each of the subsidiaries (of which Colonial Penn was one) to use the same process and quality system, in order to break down the barriers between the organizations. We put on some workshops and seminars for PACE as well, during the course of which I had the opportunity to get Dr. Deming's views on quality as a transformational system for the community. There are now almost 200 organizations similar to PACE throughout the country.

The Philadelphia Story: W.E. Deming Remembered

by Mary Ann S. Gould

At the base of the podium in a small amphitheater leaned a very old man, seemingly tired and having difficulty with his stance. He was somewhat shaky on his feet and appeared to me as a large man in a larger and somewhat rumpled suit.

This was the person a friend of mine wanted me to hear and meet? *This* was the man who could make America more competitive? *This* was the sage who could help *me* improve my business?

As I sat in the audience on that spring evening in 1981, I thought, "Deming? Who was he?" I had never heard of him. I wondered how someone so obviously drained and, I thought, possibly feeble, as well as very old—plus an academic to boot—could teach anything practical, especially for the young and rapidly growing electronics company of which I was president. Perhaps I was wasting the evening; maybe my time would be better spent meeting with key vendors at this Washington, D.C. industry trade show. Finally, he began to speak and my fears were confirmed. A wasted evening during a busy trip!

His voice was deep but strained, his words hesitant, as if he did not know where to start and was searching. His message was unclear, and I began to wonder how I could get out of there. Then, slowly, his tone deepened further and became more resonant. A sort of illumination and sparkle lit his features as the tiredness seemed to disappear and give way to a commanding presence at the podium.

And for the first time I heard his eternal challenge, *"How do you know?"* It was spoken with fervor and an almost intimidation, as if to say, "You should know, and if you don't, you better find out for that's your job!" This was followed by many questions—challenges—relating to our ways of managing our businesses and working with employees and customers.

Mary Ann S. Gould is President of the Gould Group, a Philadelphia-based total quality consulting organization. She was the founder of PACE and CEO of Janbridge, Inc. She assisted at many of Dr. Deming's four-day seminars and is a noted Fellow of the Society for the Advancement of Management.

My attention was riveted. Several points touched my own long-held beliefs. But then he would throw out a zinger, such as questioning how we managed by the numbers, etc. How in the world was I to run a company if I didn't manage by the numbers—and me with an accounting background. I kept trying to figure out what he meant. Throw out the numbers? So it went through the evening, intellectually and emotionally oscillating between points that touched my heart as well as my mind and then the outrageous and/or provoking. Thoughts and ideas seemed to be cast out as tantalizing puzzles to be picked up and put together, if only you could find all the pieces. By the end of the evening, I was angry, perplexed, curious, challenged—and hooked.

After the speech, I was introduced to Dr. Deming. Despite the fact that he truly was very tired (he had just gotten back from a trip and mentioned that he had had little sleep), he was very gracious and extremely interested in my comments and questions. In turn, he asked questions of his own regarding how I saw my job as company president. He gave me his card and asked that I telephone to talk further. Thus began a warm friendship and a sort of mentorship that was to have a profound effect on my life as well as on Janbridge and lead to the eventual creation of PACE, the Philadelphia Area Council for Excellence.

I did talk with Dr. D, studied his writings, and through his request was invited by AT&T to be their guest, along with one of my associates, at a four-day Deming seminar held in Rhode Island. There I learned even more and talked more with Dr. Deming and with many AT&T executives, managers, and line workers. For the first time, I saw some of Dr. Deming's legendary crustiness as well as his soft spot.

At lunch one day, my Janbridge associate, Harold Tassell, sat next to Dr. Deming and was asking him a pointed question when Dr. Deming suddenly just got up and left. One AT&T manager mentioned that this had happened before with a very senior executive, but not to worry since Mary Ann was here. Taking the bait, I asked why and was told I had the perfect personality—I was a pretty woman.

In a way that joke had some truth to it, for through the years it was evident that Dr. Deming enjoyed talking to women and getting that occasional hug or kiss on the cheek. Yes, he could at times truly be described as crusty or a curmudgeon, a brilliant, disciplined, funny, fervent, caring, deeply religious human who could also intimidate, infuriate, and, most of all, challenge as he taught. He was focused on a mission, and he knew the time and his time were limited.

We continued to work at Janbridge to understand and implement some of Dr. Deming's teachings. Changes began to occur. I became convinced that his message and principles were critical to the revitalization of America.

About a year and a half into working on the Deming way, I was given a special challenge to back my belief with action. As a member of the Board of Directors of the Greater Philadelphia Chamber of Commerce, I had taken part in discussions on how to attract more businesses to the Delaware Valley.

A new organizational design was developed with which I agreed but felt was not sufficient. This new plan was brought to a vote of all the directors. A large press contingent was waiting outside the meeting for the anticipated announcement. Before the vote was taken, discussion was held and I expressed my concern that we were not including efforts to help existing companies improve, and that if the goods and services of Delaware Valley firms were superior in quality, other companies would be beating down our doors to locate here.

Then the vote came. I was the only "no." I expected to be asked to resign, because this project was dear to the heart of Hal Sorgenti, President of the Arco Chemical Company. He was a person I both liked and respected, but also a very strong personality who probably had wanted to announce a unanimous vote to the press.

But I wasn't asked to resign. Instead, Hal, as I teased him later, got even with me in a different way. He said that if I felt there was a better way to help Philadelphia area companies, to go ahead and do so. He handed me the challenge of starting what was first referred to as the Business Improvement Council (BIC).

I was young enough (or was it crazy enough?) to accept the challenge, for I believed then that quality was the way. As the prerogative of the founding chairman, I began to focus the direction of BIC of the Greater Philadelphia Chamber of Commerce, which we later renamed PACE.

We had very limited resources to work with in the beginning. Joan Welsch, group vice president of the Chamber, became involved with the effort, to my everlasting gratitude, as did Bill Morlock, vice president of Philadelphia Electric Company, and at a distance—Dr. Deming.

After setting our charter to help the companies of the Delaware Valley grow and prosper through continual improvement of goods and services—to make "Made in the Delaware Valley" a mark of excellence—we then tried to figure out how we could do so on a large scale, as well as pay for it financially.

I asked Dr. Deming if he would come to Philadelphia. He said he would but his schedule for the next six months was very tight. He suggested that I talk to Bob King of GOAL, whom I had met at the AT&T conference. Bob was most helpful, both in sharing information on GOAL and arranging to have our first four-day Deming seminar in Philadelphia.

Before such a seminar could occur, we needed to drum up interest. At that time, Dr. Deming was not well know in Philadelphia, and to ask people, especially executives, who were our particular target, to give up four days for a seminar on quality was a risky venture.

On top of that, the Executive Committee of the Chamber, of which I was a member, needed to approve the expenditure of a substantial sum of money for arrangements and fees. One of the directors asked how we could be sure an eighty-two-year-old man would be able to be here to do the seminar, let alone complete four exhausting days. Someone even joked that maybe we needed a Lloyds of London policy. The Executive Committee carefully considered all points and recognized the opportunity for the Philadelphia region and eventually gave the go-ahead.

I came up with the plan of having an introduction/overview at a breakfast seminar, which actually ran three hours. Included were presentations from James K. Bakken, vice president of Ford Motor Company; Bob Cowley, general manager of AT&T Merrimack plant; myself; and, of course, Dr. Deming.

We arranged to make it a high-visibility as well as exclusive conference. CEOs and senior-level executives were formally invited by the president of the Chamber of Commerce. We also arranged for the mayor to open the meeting and for the press to be in attendance. It turned out to be a tremendous success, with over 350 top executives attending.

I have to admit now that I was scared to death. I knew from personal experience that Dr. Deming was, at least at that time, not always at his best with short presentations, primarily, I believe, because his message was needed in its entirety. When he tried to reduce it to fit a brief time, he felt constrained and frustrated and, therefore, sometimes was not received especially well by an audience, especially top executives. This was part of the reason we "sandwiched" him between two executives who could clearly demonstrate that Dr. Deming's way worked for their companies. Plus a simple fact of life—peers listen to peers.

The interest generated from the breakfast seminar plus the later publication of Mary Walton's fabulous article about Dr. Deming in the *Philadelphia Inquirer* magazine—with Dr. Deming on the cover with what I call his "Uncle Sam" look and pointing his finger—combined to produce a sell-out four-day Deming seminar. Most importantly, our goal of reaching top executives was met. Attendance was predominantly middle- and senior-level management.

Those were rough yet exciting days, preparing for the breakfast meeting and the first seminar. We had a skeleton staff. Joan Welsch was invaluable, as well as a shoulder to cry on. The continued support from Joan, Henry Reichner, executive vice president, and initially Chamber President Thatcher Longstregth and then his successor, Fred Di Bona, was critical to success. Bill Morlock and Philadelphia Electric also loaned resources to help us organize, and our BIC/PACE secretary, Rosalie DiStefano, helped us keep track of the myriad of details.

The four-day seminar came. The attendees were enthused. And the questions arose: "What next? How do we implement? Who can help us?" I wasn't sure how to answer, but I did have an idea. Since Dr. Deming believed that it was top management's primary job to bring about this change, and since top management likes to deal with other top management (the peer-to-peer grouping), why not create a learning and sharing opportunity whereby a group of Philadelphia area companies would have their top management teams come together to learn more about quality and how to lead implementation in their own firms, as well as support each other's efforts.

Dr. Deming liked the idea, and on the last day of the seminar he asked me to share it with the audience. He further indicated that he would help us where he could. Several companies expressed interest. Dr. Deming then recommended I talk with a few people who could assist with the training. Fortunately, one of those was Dr. Brian Joiner.

At first, Brian was a little reluctant, as we were taking on a major challenge and he had limited available time. Finally, we came to an agreement, and he brought in Peter Scholtes to help. I, in turn, agreed to spend time each month working with them to help develop the implementation process.

My home became the Philadelphia planning office. It was a delight as well as a learning experience to work with Brian and Peter. It was also very tough work, for these were the early days of the American quality movement. There were no detailed guides on how to bring about company-wide change, especially when dealing with top management teams simultaneously. Brian and Peter did a lot of great work back in Wisconsin, and then we would get together to review or brainstorm in preparation for the next steps. Brian and Peter provided the formal instruction at the roundtable meetings. Of course, Dr. Deming was always interested and arranged to visit the roundtable to see the interaction of companies.

Such firms as Philadelphia Electric, Rohm & Haas, Campbell Soup, the Philadelphia Navy Yard, as well as Janbridge and several others made up that first roundtable. These companies and the members of subsequent roundtables II and III provided the critical mass which was to have a major impact on Philadelphia becoming a key center for quality via the Deming Way and the growth of PACE into a nationally recognized leader in quality improvement.

Perhaps another vital turning point was our trip to Japan in 1985 to celebrate with Dr. Deming the thirty-fifth anniversary of the Deming Prize. By this time, PACE had a full-time manager, Rick Ross, who spearheaded the planning for the trip. A number of Philadelphia area executives as well as representatives from companies such as Boeing, Procter and Gamble, Ford, Hewlett Packard, etc. made up the contingent for the two-week study tour. PACE, under the auspices of the Greater Philadelphia Chamber of Commerce, sponsored and organized the delegation. Joan Welsch and Rick Ross came, as did I, Brian Joiner, Peter Scholtes, Dave and Carol Schwinn, Tim Fuller, and executives from several of the roundtable companies. Ceil Kilian, Dr. Deming's long-time jewel of a secretary, was our special guest, not only because she is such a delight but because she has always been there to help and listen as well.

Since we were associated with Dr. Deming, we were able to make special arrangements to visit a number of Japanese companies that had won or were vying for the Deming Prize. What an eye-opening education, especially since we, through Dr. Deming's request and the assistance of JUSE (Union of Japanese Scientists and Engineers), were able to meet with executives as well as managers and workers. They talked very candidly about their experiences in implementing quality improvement—what worked, what did not, where they were going next—and provided volumes of documentation.

We were also very fortunate to have Dr. Myron Tribus as our tour leader. Myron would provide background information before we visited each company as well as lead discussion summaries afterward, even when we were tired to the bone. Myron would cheerfully (some of us teased that it was really sadistically) get the conversation rolling as soon as we got back on the bus after a very full day, and he kept us talking until the last insight was expressed.

We didn't have too much time to spend with Dr. Deming since he was committed to many JUSE functions. He did, however, join the entire group for dinner one evening, and I was fortunate enough to have a breakfast meeting with him.

The highlight certainly was the award ceremony and the buffet dinner that evening. The Japanese were so polite around Dr. Deming, definitely in awe of him. Then, in we marched, hugging and kissing Dr. Deming. I think our Japanese hosts were very surprised at the demonstration of affection. I have photographs of Japanese attendees and press taking photographs of us as we embraced Dr. Deming. His pleasure—as well as the hugs—were most joyful. I am so happy we could be with him on that special day.

Dr. Deming had one major concern about our tour. Surprisingly, it related to some of the enthusiasm. We were caught up in what had been accomplished and may have become, at times, overly concentrated on techniques and total quality control methods. I personally believe our reaction was a forerunner of Dr. Deming's concerns about the TQ efforts in the United States, which in many cases lost his message and may be why he focused so much on the need for "profound knowledge" in his last years.

During a brief respite while staying at the Imperial Hotel in Tokyo, I took the time to fulfill Dr. Deming's request to provide a written summary of what we had learned. The note I got back from him is quite succinct and revealing:

21 Nov. 1985

Dear Mary Ann,

What is TQC?

Have we not much more?

I thank you for your wonderful letters.

My love,
W. Edwards Deming

Call before 9 if you can—maybe breakfast Friday.

We did talk, and talked again through the years, regarding what TQC or total quality is. I know he feared that America would be enticed with the sizzle—the techniques—and forget that all improvement need to be based on what he came to call Profound Knowledge, the path we must pursue.

That trip became a watershed, for many people on it became leaders or at least major proponents of quality the Deming way. It was also a time for me to consider a major decision. A number of companies in Philadelphia wanted to join a future roundtable, but Dr. Joiner's schedule did not permit taking on another group. I talked with Dr. Deming, and he encouraged me to work with the next group of roundtable companies, with the provision that I include a statistician, because this was not my field. Thus came a partnership with Ron Moen, a true Deming disciple, to lead a second roundtable. Later, I also led a third roundtable with the staff of the Gould Group and assistance from Tom Murtha and Dr. Jim Donald and support from Dr. Dick DeCosmo, president of Delaware Valley Community College.

I loved—and at times hated—the roundtable work, for it was difficult working simultaneously with groups of top-level executives. Additionally, they expected absolute polish and all the answers while we were still learning what implementation was all about. Yet at the same time, we were forging new paths and developing new processes. It was frustrating at times but worth it personally, because I learned—had to learn—so much and had the opportunity to develop materials which are in common use today. Also, as previously stated, many of those participating organizations became major leaders in quality and the Deming way.

One true regret was that as a consultant I could no longer chair PACE, an organization that will always be very dear to me. PACE was growing, and we hired a director, Maureen Glassman, who proved to be a great asset in expanding PACE's curriculum and organizational participation while upholding the mission. Maureen also proved to be a wonderful spokesperson for PACE. She came to share a great admiration for Dr. Deming and spearheaded, along with Nancy Brout, the Deming Philadelphia Seminar Organization, and always enjoyed the staff night out with the master. Our association with Dr. Deming deepened, and in 1986 he promised me that he would try to set aside one seminar a year for PACE. We in turn promised to use the funds generated to further PACE's mission.

Personally, I was profoundly impacted by my long association with Dr. Deming. I attended numerous four-day seminars as an assistant, leader, and speaker. Each time I learned more, but perhaps most personally memorable were the visits as well as phone calls and letters received through the years.

Dr. Deming had sort of an Old World manner. I'll never forget the phone calls he would make to my home. I would answer the phone and invariably hear his deep voice asking, "Is this 'Miss' Mary Ann?" Sometimes talking on the phone was difficult, since he had a hearing problem and I have a very slight hearing impairment; the results were at times confusing and/or hilarious.

I most enjoyed meeting with Dr. Deming in Washington, D.C. where, when not under the strain of seminars, he relaxed and became even more jovial and wide ranging in his conversation. He especially liked to have dinner at the Cosmos Club, and he always made sure I ordered hazelnut ice cream for dessert so he could eat his—and most of mine, too.

For many years he would insist on driving his well-preserved 1969 Lincoln Continental. Some of the rides were a little scary, as driving was not one of his greatest skills, possibly due to eyesight problems.

On one particular visit, I was accompanied by Barbara Lewis, an associate of the Gould Group and the person who helped Dr. Deming during his Philadelphia visits. We had dinner with him on a Saturday evening—of course at the Cosmos Club—and then spent some time reviewing Gould Group work. Dr. D asked us if we would like to attend church with him on Sunday, before which he would make us breakfast at his home.

Barbara brought along a camera the next morning, and we took some photographs, which I cherish as a unique expression of the Deming message as well as the man. Dr. Deming prepared the entire breakfast himself and would not let us help because we were his guests. He also was very proud of the spoonbread he made as one of the entrees.

To look at the photographs today is both funny and thought provoking. Here is a world-famous man, wearing oven gloves, bending over a stove, with full concentration on what he is doing. It is as if doing this well, in order to give his guests pleasure, is the single most important thing in the world. His facial expression and body language of full concentration on the job at hand is edifying.

He also loved music, especially the Gregorian Chant. He was very proud of the choir at his church, and I believe he had donated the organ in addition to contributing to major roof repairs, etc. As with cooking breakfast, Dr. Deming at prayer was fully in the prayer, fully with his God, offering his thanks and praying very deeply.

Through the years, I received letters or postcards, especially when he was visiting a new country. His letters were always very gracious and often extravagantly complimentary. He could made you feel very special and was thankful for the least little thing. He would usually handwrite the correspondence, using dark black ink in a striking script.

> ...I owe you letters, we had a good chat on the telephone didn't we?...I admire you, with affection...
>
> My Dear Mary Ann, I would do anything for you. Your success in Philadelphia is phenomenal...
>
> [Postcard from Westminster Abbey] Am giving 4 day seminar here. Great! With admiration & affection, W.E.D.
>
> ...It was a delight to see you at my class" [NYU]...I am wondering if you would be good enough to talk to my class on 18th February...
>
> ...It was good to see you in Philadelphia. You are wonderful. I think of you whenever I think of Philadelphia. You are a great leader...I send my best thanks...
>
> ...It is always a joy to work with you...
>
> This is just a note to tell you that I am thinking of you, wish for you all that is good...With affection, W. Edwards Deming
>
> [Postcard from New Zealand] I wish that you were here to help top mgt. to understand their job...
>
> ...The seminars exhaust me...

Once, after I had been ill, I received a letter in which Dr. Deming profusely apologized that he had not been as up to date in his calls as he would like and had not known of my illness but that I was in his prayers.

He was also very helpful, always interested in the work I was doing as well as the companies involved. For example, several executives and managers from a large Texas company where I was consulting were to attend a Deming seminar in the D.C. area. I mentioned this in passing to Dr. Deming, asking if he might talk with them for just a few minutes at the end of one of the sessions. Dr. Deming graciously said yes and went further, asking if I wanted to join him so we could meet with the group together.

It turned out that we spent several hours with Dr. Deming. The executives were captivated by his wit as well as his comments and questions relating to their business. It was a wonderful evening. The important point, though, is that here was a man who was clearly very tired and who had much work to do, yet a simple request for a few minutes turned into so much more. And therein is the story of his life and work.

Incidentally, when I was having breakfast the next morning before returning to Philadelphia, I received a message that Dr. Deming wanted me to speak to the seminar audience—in fifteen minutes. This, too, was Dr. Deming, for he expected you to know your job.

He also loved to hear stories that demonstrated learning. This was especially so if the examples were funny and involved results from his cherished "willing workers" viewpoint. I remember one conversation where I described a skit that an AFIDCO, Inc. (Nestle subsidiary) work team did to demonstrate the "old" ways. The team had put together a very comical (with a lot of underlying truth) play, ending with the production of a pizza instead of the expected product. They then demonstrated the changes due to employing the Deming principles and the integrated vision implementation process we had employed. The operators had written and performed the entire skit, and it was so apt. Dr. Deming just roared with laughter—his deep "guffaw"—and asked for a copy of the video. He gave me a message of appreciation to the team and management for a job well done, which meant a great deal to them.

Perhaps one regret I have is that over the last two years I became so involved in some research and developmental work on managing change and holodynamic organizational structure that I did not get to talk with Dr. Deming as much as before. Yet when we did talk or meet, he was most supportive, always interested, and always encouraging.

I had given him some outlines relating to use of "boundaries" in understanding the effect of changes in organizations and mentioned that this was part of the direction I was taking, working with Dr. Shaile Stephens. He immediately asked questions of Shaile's background and said that we should all get together. Unfortunately, this was not to be.

I have to confess, though, that this was in part due to my own trepidation. For as much as I adored Dr. Deming, I had a reluctance to show him work unless it was very complete. Perhaps this was due in part to my own desire to meet his expectations and in part because I had been on the receiving end (on more than a few occasions) of his, shall we say, "dissatisfaction."

He never really said anything that could be taken as harsh criticism. There would be a question or a comment and then a silence. Dr. Deming's silence could be very loud. He made you rethink. Sometimes I would change because of those silences and sometimes not. Yet he always respected my right to pursue my own search for "profound knowledge."

Thus, the legacy I have from him is not so much the letters, postcards, pictures, advice, and friendship, but primarily an ongoing living challenge, a challenge issued on that long ago spring evening and reinforced through the years: *"How do you know?"* It is a question that continually drives me to learn and to grow. As Marcel Proust wrote, "The real act of discovery is not in finding new lands but in seeing with new eyes." Dr. Deming helped me to see with new eyes and left me, along with all of us, with the challenge to learn, to grow, to become more of who and what we can be.

His death tremendously saddened me. Since my own father was gravely ill, continually in and out of intensive care, I was unable to go to Washington to pay my respect. I did, though, go to my own church to light a candle and leave a rose in memory of a special and unforgettable teacher and friend.

Thank you, Dr. Deming, with my affection and admiration always! Thank you for teaching me to know how to know.

A Garden of Memories for Dr. W. Edwards Deming

by Maureen Glassman

In 1983, the Philadelphia Area Council for Excellence (PACE) was created by a visionary group of volunteers and staff of the Greater Philadelphia Chamber of Commerce. Its aim was to assist local businesses to survive and thrive in the face of increasing competitive pressures and a marked economic downturn.

During the early 1980s, hundreds of Philadelphia manufacturers had moved south, moved offshore, or closed their doors completely. In response, chamber volunteers and staff began researching what the chamber could do to help. Mary Ann Gould, a volunteer leader and business owner, attended a four-day Deming Seminar in 1982 and brought, with some trepidation, Dr. Deming's revolutionary message back to her colleagues at the chamber.

With early help from Bob King of GOAL/QPC and Dr. Myron Tribus, one of the first regional excellence councils in the nation was born. Our plan was to use the management philosophy of Dr. W. Edwards Deming to transform the Delaware Valley. The original vision/mission of PACE was "to establish the Delaware Valley as a model region known for the quality of its goods and services and the productivity of its workers. In acquiring this, we enhance the competitive edge of local companies and thereby strengthen the region's economy."

In 1983, the cost of producing a four-day seminar for several hundred people was mind-boggling and quite a financial risk for a chamber of commerce. In addition, in the early 1980s Dr. Deming's name was just beginning to be known in the United States. However, then Chamber President G. Fred Di Bona and Executive Vice-President Henry Reichner, prodded by Group Vice-President Joan Welsch, had the courage to take the leap of faith and support PACE in its efforts to bring Dr. Deming's message to our region. Thus began

Maureen Glassman is Executive Director of the Philadelphia Area Council for Excellence (PACE). She has been a Deming disciple since the early 1980s.

a ten-year relationship which included Dr. Deming's return to Philadelphia each year to revitalize and inspire PACE companies that were working to understand and implement his philosophy.

As director of PACE, I had the privilege of hosting Dr. W. Edwards Deming's annual seminar in Philadelphia from 1986 to 1993. Each year I marveled at his stamina, his sense of humor, and his absolute belief in the innate goodness and integrity of people. A consummate teacher and learner, Dr. Deming modeled continual improvement in his seminars. He constantly revised his notes, added cogent stories and illustrations to his lectures, and delighted in incorporating new insights and examples supplied by his colleagues, seminar helpers, and students.

In 1986, PACE was fortunate to be sponsoring the Deming Seminar the week before Dr. Deming's eighty-sixth birthday on October 14. On that Wednesday in October, Dr. Deming conducted his famous Red Bead Experiment for over 400 participants from 100 companies, colleges, and government agencies. After he had completed the experiment, PACE staff and twenty volunteers unveiled a surprise birthday party, to the delight of the crowd and the initial puzzlement of Dr. Deming. This special event featured a red bead simulation with over 2000 red and white balloons, an enormous paddle to capture the balloons, a "boss" construction hat for Dr. Deming from one of the attending companies, a giant birthday cake, and even a Philadelphia Mummer's String Band for entertainment.

As the energized audience rose and joined seminar leaders Ron Moen, Mary Ann Gould, and me in a thunderous rendition of "Happy Birthday, Dr. Deming," I was thrilled that we were able to express our affection for him in such a unique way. The party planners really took "joy in their work." Happily, Dr. Deming's pleasure was obvious as he smiled, willingly posed for photos, and cut his cake. Soon, however, and true to form, he turned to me and said, "That was wonderful! Now it's time to go back to work!" Then, while the audience munched on his cake, thousands of red and white balloons hovered overhead, and we scurried to clear the stage of party clutter, Dr. Deming calmly continued his seminar.

On another occasion, our leading utility and PACE founding member company, Philadelphia Electric Company, helped us to thank Dr. Deming for returning to Philadelphia. Displayed in lights on their corporate headquarters high above the city, PECO spelled out "Philadelphia Welcomes Dr. W. Edwards Deming." Picking him up at the Amtrak Thirtieth Street station was ideal that evening in February 1989, because the building with its special message was directly across the street. Dr. Deming smiled, squeezed my hand, and said simply and graciously, "Thank you, Maureen. It's good to be back in Philadelphia."

According to Ceil and his daughters, Dr. Deming very much enjoyed his annual trek to Philadelphia. I particularly treasure the memories of our Thursday evening dinners in the four-star restaurant at our hotel. For the first five years, only PACE staff (all women), Jan Gaudin of Boeing Helicopters, and Philadelphia journalist Mary Walton, author of *The Deming Management Method,* were allowed to attend. In 1991, we were finally able to invite John Ely and his helpers from Markeys' Audio Visual Company, who traveled with Dr. Deming and were sole suppliers for audiovisual equipment and services for all his four-day seminars. John and his associates were honored to be included in the last evening's unwinding and celebration. Again, thanks to Dr. Deming's introduction, PACE continues to employ Markeys' as our sole supplier for all major quality events because they are truly dedicated to delighting the customer.

Dr. Deming returned to the Delaware Valley eleven times to support PACE's efforts in creating a community of excellence. He believed that we were building a "critical mass" of people and companies who were committed to learning a new and better way to manage our organizations. Early on, after his first visit to Philadelphia in 1984, he asked his former student and colleague Dr. Brian Joiner to come to Philadelphia and work intensively with the group of nine companies that made up the first PACE Roundtable.

The first PACE Quality Roundtable companies, including larger firms such as Rohm and Haas, Campbell Soup Company, and Philadelphia Electric Company, were truly quality pioneers. They were learning a completely new way to manage and supported each other in uncharted waters. Few roadmaps and guidelines for transformation existed in 1984 when the roundtables began.

At Dr. Deming's memorial service in Washington on December 29, 1993, Peter Scholtes reminded me that he was first hired by Brian to provide organizational development assistance to the roundtable companies. Even as early as 1984, Joiner Associates recognized that Dr. Deming's philosophy required radical cultural change in addition to systems thinking and a deep understanding of common cause and special cause variation. For nine years, Dr. Brian Joiner and Peter

Scholtes have been friends and supporters of PACE, and we have Dr. Deming to thank for our first introduction to these wonderful quality leaders.

In subsequent years, PACE initiated Roundtable II, Roundtable III, and a Quality Forum led by associates of Dr. Deming, including Ron Moen and Dr. Tom Nolan and local quality consultants such as Mary Ann Gould and Dr. Bonnie Kay. The local community college, Delaware County Community College (DCCC), was also an active player. Along with management teams from several local companies, DCCC's president and executive staff attended the PACE Roundtable II. In intensive monthly training sessions, they learned Dr. Deming's theory of management. Subsequently, college staff acted as facilitators, assisting the consultants who led Roundtable III.

Dr. Richard De Cosmo, president of DCCC in Medin, Pennsylvania, was the leader of the college team that attended the PACE Roundtable meetings in 1986–87. Strongly influenced by Dr. Deming's message, Dick De Cosmo became one of the first college presidents in the United States to fully commit his institution to total quality transformation. Today DCCC is a national leader among educational institutions, providing excellent total quality management training for local organizations and breaking new ground as it continues to implement total quality in administration, curriculum, and teaching processes.

Additionally, DCCC became the first educational institution (currently there are nine colleges and universities) to join PACE as a charter member, participating actively in PACE learning opportunities as well as working with us on special efforts to promote total quality in our community. For example, the college provided quality project team training for several area hospitals in conjunction with PACE and supported by a generous grant from the Pennsylvania Ben Franklin Partnership Program.

Over the years, this start-up effort developed into the Greater Philadelphia TQM in Heathcare Network, a learning consortia of twenty-six regional hospitals dedicated to "improving the quality of patient care through the application of total quality principles." Coordinated by PACE, this unique project fosters sharing and learning about total quality in a critically important sector of our community.

PACE's outreach to the community extends well beyond healthcare and education. Member organizations from these sectors as well as federal and local government, manufacturing, service, and small business acquire new knowledge and share resources and experiences with each other through the bimonthly Quality Network meetings, the Annual Quality Conference, and monthly quality management seminars. Over 500 corporate and individual members and 3500 people annually participate in PACE programs and services. In addition, over forty quality leaders from all sectors of the community serve on five PACE teams which provide direction, resources, and the "voice of the customer" to PACE staff.

As PACE progressed in its ten-year quality journey, our volunteers and staff revisited the vision and mission several times. Today, we continue the thrust of our original vision: to create a community of excellence in the Delaware Valley that will stimulate the regional economy and improve the quality of life for all.

Our mission has evolved, however, to emphasize both PACE's critical role in promoting a learning community and Dr. Deming's teaching that experience without theory teaches nothing. It reads: "PACE will promote and provide opportunities for the transfer of knowledge about Total Quality, based on theory and practical application, that will further organizational improvement and growth." During the next decade, we will doubtless revisit and revise our mission, yet again wrestling with its meaning for our organization and working to ensure that it responds to the changing realities and needs of our community, members, and volunteers.

Influenced by Dr. Deming's philosophy of inclusion and partnership and wanting to accelerate the quality initiative in our region, PACE joined last year with three local community colleges and two economic development organizations to form the Delaware Valley Total Quality Consortium (DVTQC). Its mission is, "To work in partnership with its customers to promote and facilitate the implementation of TQM in all sectors. This will create a culture of teamwork and cooperation to improve quality, productivity and competitiveness throughout the region."

The DVTQC jointly sponsors special projects designed to leverage our individual resources as well as access state economic development funds. These projects require collaborative efforts of many groups and could not be done as successfully or effectively alone. Initial efforts include the TQM Regional Programs Calendar (which lists over seventy quality courses presented by eleven local institutions or associations and is published biannually), the Delaware

Valley Quality Recognition Process (non-competitive process that includes a recognition luncheon and an annual publication which documents quality success stories and lessons learned based on the seven Baldrige criteria), and the Total Quality Customer Survey (in-depth study, quality achievements, and current challenges of 200 local companies).

As PACE develops new partnerships for quality, responds to customers' needs, and promotes the quality movement in the Delaware Valley, we continue to be influenced and enriched by Dr. Deming's example, philosophy, and vision. To recognize his contribution to our community and to create a living tribute, in 1991 PACE initiated a special fund to restore the 100-year-old Japanese Glen in Fairmount Park in his honor. Begun with a generous donation from PACE's parent organization, The Greater Philadelphia Chamber of Commerce, the Deming Glen Fund has received contributions from many individuals and organizations. Dr. Deming's daughter Linda and his long-time aide and secretary, Ceil Kilian, have agreed to participate in the dedication ceremony and contributors' reception once the project is completed.

Dr. Deming's contribution to the Philadelphia community, to the nation, and to the world is immeasurable. Together with the PACE staff, volunteer leaders from our member companies, and all those who were touched by his profound message, I mourn his loss and celebrate his achievements. Personally, I have been deeply influenced by Dr. Deming's management philosophy and have tried to incorporate its concepts into both my personal and professional life. His legacy to all of us is a vision of a new world in which people, communities, governments, and nations recognize their interdependence and work together to achieve personal and organizational transformation.

Dr. Deming traditionally ended his PACE seminars by simply saying, "I have done my best." It is our challenge and joy to continue his work and do likewise.

The biggest problem that any company in the Western world faces is not its competitors, nor the Japanese. The biggest problems are self-inflicted, created right at home by management that are off course in the competitive world of today.

Everyone doing his best is not the answer. Everyone is doing his best. It is necessary that people understand the reason for the changes that are necessary. There is no substitute for knowledge.

W. Edwards Deming
Executive Excellence
February 1987

Chapter Eleven

DRIVE OUT FEAR

"Drive out fear so that everyone may work effectively for the company."

Dr. Deming often spoke about the need to be free from fear and to feel secure. He points out that *secure* comes from the Latin *se,* meaning without, and *cure,* which means fear. Being secure means being unafraid to express ideas, to ask questions, or to point out problems. He said that the "common denominator of fear in any form is impaired performance and padded figures." He also talked about people being afraid of knowledge, partly due to pride and partly due to a lack of money.

Dr. Deming often spoke about the economic cost of fear, which he said is staggering. He told the story of a young man who did not understand his job but was ashamed to admit it because he did not want to appear slow or incapable. He was afraid to point out problems because he did not want to be labeled a "whistle-blower," so he kept his mouth shut and the plant went out of business. "Could he have saved the plant?" thundered Dr. Deming. "Who knows? We will never know because the system failed us once again"—in its failure to facilitate the upward flow of bad news.

This was one of the big problems in the NASA program and has continued to plague it to this day. One night, Myron Tribus and Dr. Deming got into a discussion about what happened at Cape Canaveral to cause the Challenger accident. I happened to be at the table with them and had recently made a presentation in the United Kingdom that had involved the Challenger story. At one point, Myron turned to me and said that since I had researched the Challenger accident, perhaps I could shed some light on the subject. I swallowed once or twice, took a few gulps from a glass of water, and told him the burnt toast story, a story that Marshall MacDonald told about NASA and America back in 1981. When they made toast and it burned, they and scraped it clean instead of fixing the toaster. Dr. Deming roared, "that's because they were afraid of getting caught with burnt toast in their hands."

His point was right on the money, as usual. Good management will solicit recommendations from employees, act on the good ones, and describe in detail the reasons for not acting on the others, thereby earning the respect of the workers. Dr. Deming taught that mutual respect among employees and management is essential in order to eliminate fear and promote constructive activity. Another loss due to fear, he pointed out, is the inability to serve the best interests of the organization because of the need to satisfy the rules or meet a deadline.

Fear operates in the dark and in the shadows. The basis of its power is confusion and mistrust. J.A. Frounde once said that "Fear is the parent of cruelty." In the old English form, the word was the verb *fieran,* which meant to terrify or to take by surprise. In the noun form, it appears as fright, dread, terror, horror, panic, alarm, distress, consternation, and trepidation. Adolph Hitler once called overpowering fear "the weapon which most conquers reason, terror and violence." Panic is a sudden frantic fear, often groundless, while alarm is fright aroused by the first realization of danger. Dismay is the apprehension that robs one of courage and of the power to act effectively, while consternation is a state of often paralyzing dismay that is characterized by confusion and helplessness. Trepidation is a feeling of dread, marked by trembling and hesitancy to act.* All of these words describe fear in its many stages and forms, and often Dr. Deming portrayed it in a progressive sense.

Dr. Deming also pointed out that some managers believe that a certain amount of fear is necessary to get the work done. The real issue here, he said, was the failure to separate tension from fear. While it is true that a certain amount of tension is positive and healthy, because it gets the adrenaline flowing, too much is destructive because it leads to burnout and fear. He believed that fear among

* *The American Heritage Dictionary of the English Language,* 3rd edition, 1992.

salaried workers was largely propagated and intensified by annual performance reviews.

He believed in learning from the workers, as he continually points out in his books and writings. Remarks similar to the following were made to him constantly by the rank-and-file: "I am afraid to admit I made a mistake," "My boss believes in fear," "He can't manage his people if they don't hold him in awe," Management is punitive," and the classic "We mistrust the management. We can't believe their answers when we ask why we have to do it this way."

He also believed in learning from his students as much as they learn from him. The following contribution is by Lisa McNary, one of a class, a class of his last graduate students. She has a message to offer, and I believe we all can learn from it. The message is about candles that light both the past and the future. We believe Dr. Deming's teachings were candles of quality for all time, with a hunger and thirst for knowledge, the main weapon in driving out fear.

On Quality Management, Dr. Deming, And Candles: The Last Graduate Student Remembers Her Mentor

by Lisa McNary

It's rather pleasant the way the human mind slips backwards and forwards through the years. Looking back through the years can be rather like walking down a corridor holding a candle. Incidents and places completely forgotten appear out of the blackness and, one by one, are lit as you pass.

Jean Hersey
The Shape of a Year
(Charles Scribner & Sons, 1967)

Lisa McNary is one of Dr. Deming's last graduate students. She received her doctorate from the University of New Mexico.

At Dr. Deming's passing, I looked over the internship journal that I kept as a requirement for a management course while working on my doctorate. In that internship, I spent several weeks with Dr. Deming, working with him in Washington and at both New York University and Columbia University. In a few short months, the words he spoke to me in reply to my inquiry regarding his sponsoring my internship were true: "Spend some time with me. You'll learn more than if you did an internship in a company." I did indeed. I learned a lot about quality management, I learned a lot about Dr. Deming, and I learned a lot about life. I only wish there had been more time to learn than the three semesters that I spent with him. I have the dubious distinction of being Dr. Deming's last graduate student. In many ways, the candle that lights the memories of the past is the same candle that must be carried into the future, and so Dr. Deming's work continues.

My relationship with Dr. Deming began in the winter of 1992. I was nearing the dreaded ABD (All But Dissertation) stage of my doctoral coursework, that dubious point where some graduate students begin their research, only to get mired in the process for years. During that semester, I took a research methods class that allowed me to pursue dissertation ideas; it was a conscious attempt to keep me on track to graduation. Several ideas popped up during my varied coursework, but none really grabbed me as "the one topic." Then, information regarding Dr. Deming's System of Profound Knowledge, which was unpublished at the time, was passed on to me by Howard Gitlow at the University of Miami. We bounced some ideas around, and I wrote to Dr. Deming and requested a meeting to discuss the topic with him, only half expecting that he would have the time or inclination. Within days, I was excited to receive a reply with a confirmed meeting date set for Thursday, March 19, 1992 at his home in Washington, D.C.

I thought that I had come prepared, with a tape recorder, my notes on quality management which I had studied since 1986, structured questions, and the first draft of a profile inventory based on the System of Profound Knowledge that would identify Deming and non-Deming managers, which was the focus of my research. Dr. Deming patiently answered my questions and advised me on some areas of additional study. Although there was much work yet to be done, I felt confident that the profile development was feasible.

Then, just before we left for dinner at the Cosmos Club, Dr.

Deming took all my materials away from me and asked, "Now, why are you really doing this? What's your aim?" This is the point at which I realized that my outward preparedness of tape recorder, notes, and questions did not equal total preparedness. I recall that I needed my notes desperately, since they contained my rationale for undertaking as a topic the System of Profound Knowledge, which was still very new to me. I struggled with his questions until he was reasonably satisfied that I was marginally competent to embark on this project. The repeated phrase "I need to study further" sufficed for now, but I knew that the time would come when Dr. Deming would expect me to have studied. I vowed to follow through.

As we trekked to the Cosmos Club in the blue Maverick, I had just enough time to realize that completing this project properly would be impossible without Dr. Deming's assistance. Dinner seemed extremely short as I pondered how to approach the topic of asking one of the busiest men in America to consider working with me on my dissertation. Over dessert, I asked Dr. Deming if he felt the project had merit, to which he replied, "You have much to study." Continuing, I inquired, "If I study and work on the profile extensively, would you be a member of my dissertation committee?" He tilted his head toward his right shoulder slightly, smiled broadly, and spoke one word that was magic to my ears: "Certainly." Thus began a series of trips from the University of New Mexico to the East Coast to meet with Dr. Deming as I finished my graduate work.

We arranged our next meeting for August 16 and 17. It seemed like a long time away, but it really wasn't. Before our meeting, my first hurdle was to pass my doctoral comprehensive exams, which were scheduled in mid-July. That gave me only one full month to write my dissertation proposal, on which I wanted to have Dr. Deming's approval before I handed it in to my department and the rest of my committee. It was during this August meeting that I realized that Dr. Deming had a far different teaching style than any professor I had ever had. Throughout his reading, he commented that the information in this draft was markedly improved, and I would beam inside at those words. Then, he intermittently sharpened his pencil every few pages to make notes and grammatical corrections, and I became dismayed. I vividly recall thinking, "How can it be good if he's sharpening his pencil so much?" When he handed the paper back to me, many sentences and paragraphs had been completely scratched through, with the word "no" written next to them or a comment such as "pronoun without an antecedent" written in the margin. Dr. Deming's style was very Aristotelian in nature in that he never specifically told me what to do, but only that certain areas needed more "study." It was up to me to find my own way to formulate questions to ask him for direction and then to study the materials once again. At first, this style greatly frustrated me. Previous professors had always given very directive suggestions, such as citing particular sources, including additional topics, or correcting my style with their own phraseology. All that was required was to incorporate those suggestions.

Dr. Deming's method required a great deal of time and effort and thus was not the most efficient way for a time-deprived graduate student. In the end, though, his method was the most effective way for me to truly learn the topic, because it forced me to read and reread, think and rethink, write and rewrite. My dissertation research became an applied version of the PDSA (Plan, Do, Study, Act) Cycle.

It wasn't all hard work, however. Dr. Deming's ideology was also that hard work deserved a bit of respite, whether it be a good meal, a classic piece of music, or a drink of gin and a slice of cheese before bed. We had pre-arranged times when we would stop our work for the day. Usually it was late, but I welcomed the time to linger at the end of the day, because this is when I learned the most about Dr. Deming as a person. Perhaps my hailing from the West stirred memories of his boyhood days in Wyoming, from which he would share anecdotes. For example, when Dr. Deming met my friendly Russian Blue feline, Master Barne Taggart Esquire, he recounted his memories of a cat he had as a boy that kept his brother and him warm during the long, cold winter months.

In fact, these personal moments contrasted greatly from many of Dr. Deming's public moments, when he would rail against unprepared reporters, errant CEOs, and those seminar participants who attempted to challenge the quality management principles. Many people asked me if he was as obstinate and rude to me as he was in those notorious railings, especially after the "American Interests" segment was aired on PBS in the fall of 1992. "Never," was always my emphatic reply. I recalled these questions and Dr. Deming's public persona as his large frame bent over to reach out to pet the entranced feline, and he cooed softly, "Come here, kitty kitty." Those two were instant friends, and Dr. Deming's first words to me the next morning were, "The kitty slept with me all night long." He smiled broadly as he stroked Master Barne Taggart Esquire, who was perched contentedly on his lap. After the cat did a series of sit, sit-up, and jump tricks for Dr. Deming, he proclaimed him "a quality cat," a badge of honor for the once stray feline.

Another memorable feature of Dr. Deming was his humor and wit. Anyone who ever attended one of his seminars had to wonder if he ever harbored a desire to be a stand-up comic. Humor came so naturally for him, and it held the attention of his audiences. When Dr. Deming gave a public lecture sponsored by the University of New Mexico in Albuquerque after my dissertation defense, I counted no less than two dozen times when he had the near-capacity crowd at the convention center in stitches during a two-hour period. One audience member asked, "What advice do you give a twenty-four-year-old graduate MBA student beginning to take on the business world?" Dr. Deming replied without hesitating, "Well, it's a little late to give him advice. He's already been to school...A little late. Yep. Another one [question]?"

His style cast a spell, with an interesting mix of his quality management theory, pertinent anecdotes, and humor. I remember being amazed that from the stage I could hear papers shuffling in the back row as one participant intermittently took notes in a frenetic manner. Nearly two thousand people sat spellbound, and Dr. Deming later commented what a good audience they were. He humbly took the compliment that it was he who held the audience. This is not to say that his style was easy. In fact, several people remarked that listening to Dr. Deming took a lot of work and concentration; however, many people also noted that they got much out of his lecture.

Dr. Deming's humor and wit often were funny ways of being serious. I recall us setting a date for my doctoral defense while at a seminar in St. Louis in October of 1992. We matched–merged our open dates, and he seized April 1: "April Fool's Day. Now that does seem to be a perfect date for a doctoral defense, doesn't it?" Even as an academic for many years, Dr. Deming realized that many of the procedures and motions that students had to go through were frustrating at best and meaningless at worst. He also knew that they were "systems" problems over which the student had no control, and thus he cooperated to get the student through the maze. Another example of his using humor to make a serious point also comes from the public lecture in Albuquerque. In making a point about the destructive effects of competition on business and the consumer, he commented:

> I'd like to understand this audience a little better. I wonder how many people here have ridden in an airplane from one city to another during the past three months, six months? Hands up. [He scans the audience.] Mmmn.. Everybody. That's what you get for competition. Could it be worse? Wait a month! It will be worse!
>
> If that airplane comes in late, the flight attendant thanks the people, thanks the passengers for their cooperation. Well, what else can prisoners do? Prisoners are pretty cooperative. Open-ended competition—as the result of which you have no choice. Northwest Airlines is the only one from Washington to Detroit. And when it arrives, when the airplane arrives in Washington or Detroit, the flight attendant thanks the people, thanks the passengers for choosing Northwest Airlines. How the hell else did they think they'd get there? They don't have a choice!

I learned a lot working with Dr. Deming, but I also remember laughing a lot. That combination of continuing to learn and laughing must have been one of Dr. Deming's secrets to a long life. If I had a dollar for every time I was asked, "How did you manage to get Dr. Deming on your dissertation committee?"—all my graduate work would have been fully financed. No one ever believed my pedestrian answer, "I just asked him." Everyone expected a long, drawn-out, amazing story. That Dr. Deming, so famous a person, would respond to an ordinary request from an anonymous graduate student just boggled their minds. At first, it boggled my mind too. I knew that I was extremely fortunate to learn from and be associated with Dr. Deming. It was his ordinary acts, however, that made him so extraordinary as an individual to me. Through all of his fame, he somehow managed to maintain a center, a balance to his life, something that I imagine to be very difficult. To me, Dr. Deming is the only person in my life who I can characterize as being a gentleman and a scholar. Many people attempt this admirable goal, but most only manage to approximate it.

Now, I am at the end of my personal contact with Dr. Deming, but I will always have contact with his work. During my doctoral work, I managed to incorporate the study of quality management into many classes, including history, American studies, and organizational behavior. Looking at the topic from a variety of lenses has given me the opportunity to turn the topic inside out and finally begin to understand its applications to work specifically and life in general. I realize that I still have much to learn. Though our lives intersected only for a short time, Dr. Deming touched my life so profoundly that it is difficult to articulate. At times, I find myself returning to the following words of Dr. Deming's, which I recorded during my first four-day seminar on "Quality, Productivity, and Competitive Position" held in Newport Beach, California on February 24 to 28, 1986, a seminar during which I was the "Recorder of Red Beads" for his now famous Red Bead Experiment:

> Continual improvement allows people pride with increased productivity. But remember there is NO instant pudding. It is a long journey. Don't tell me ten ways I can't do something; tell me one way I can! It's so easy to do nothing! It's a challenge to do something. Learning is not compulsory; it's voluntary. Improvement is not compulsory; it's voluntary. But to survive, we must learn. The penalty for ignorance is that you get beat up. There is no substitute for knowledge. Yet time is of the essence.

These words certainly haunted me for years and in many ways created the matrix around which many activities in my life have been shaped. Now, the words will continue to haunt me as Dr. Deming's last graduate student, but it also my responsibility to make those words haunt others, such as my students, colleagues, and clients. Dr. Deming is now in that "other," more perfect world. Yet while in this world, he did much to make organizational America, educational systems, and governmental institutions better. The candle mentioned in the opening quote of this essay not only lights the corridor of past memories, but also shines down a long corridor of opportunity for all of us who knew Dr. Deming and heard his words to continue his paradigmatic, landmark work. We must continue to "find the way," the better way—the Deming Way.

"You have heard the words; you must find the way. It will never be perfect. Perfection is not for this world; it is for some other world. I hope what you have heard here today will haunt you the rest of your life. Then I have done my best."

W. Edwards Deming
Deming Management Seminar
Newport Beach, California
February 24 to 28, 1986

Chapter Twelve

ELIMINATE SLOGANS, TARGETS, AND PRIZES

"Eliminate slogans, exhortations, targets and prizes for the work force which ask for zero defects and new levels of quality and productivity."

Dr. Deming believed with all his heart that slogans, banners, posters, and pledge cards do not help people do their jobs better. He said that most people already want to be proud of their work and only need the tools, methods, and organizational culture to help them do their jobs better. The problem with slogans, he taught, is that they often cause resentment and skepticism. They sound good and are catchy, but often are hollow-sounding phrases when confronted with reality.

He also disliked prizes and awards and often spoke of the U.S. Malcolm Baldrige Award in unkind words. At the PACE conference in 1986, he criticized me for using the favorite expression of then Florida Power and Light Chairman Marshall MacDonald: "Do it right the first time." "How can the worker possibly do it right when the specs are off, the machines not in good working order, and the training insufficient?" he asked. He thought it to be nothing more than words if management fails to provide the means to the end it proclaims to desire. I was beginning to understand, however slowly, what quality management was all about.

Actually, my introduction to the Deming brand of quality first occurred in the mid-1960s, when I was a student at St. John's University in New York. My mentor in statistics at the time was Dr. Jean Namais, a good teacher and a real stickler for details. She brought Dr. Deming in for a lecture on statistics. It was so deep and profound that I had no idea what he was talking about, and neither did anyone else, except for Dr. Namais. Afterward, she took great pains to explain to those of us who cared to listen each detail of Dr. Deming's lecture. She told us that he was a genius and that, like Einstein, felt that if education was to be of any value at all, it must perpetually challenge accepted meanings. Dr. Namais explained Dr. Deming's view on education, which was to create situations in which students can question officially certified facts. We learned that through these challenges, students and teachers gain the ground from which new questions are cultivated, and through new questions, new ways of seeing and learning are developed. The ultimate result of these insights, argued Dr. Deming, leads to breakthroughs and an awareness of relationships never seen before. This is what it means to be in the presence of a genius.

During the next thirty years, I would come to understand more about the concept of genius, as I had the opportunity to interact with some who epitomize the word. Genius is a higher order intelligence that involves the ability to see connections among a wide variety of events, the ability to see beyond, where no one else has seen before, and to see possibilities, even if they make others uncomfortable and seem bizarre at the time. In 1975, while I was lecturing at New York University on the theme of productivity and efficiency, I again encountered the genius of Dr. Deming. He called me aside and asked me what I did for a living. I told him that I was an industrial engineer, an efficiency expert—to which he replied that an efficiency expert was an oxymoron. At first I thought he was calling me a moron. I must have looked surprised because he explained to me that an efficiency expert was neither efficient nor an expert. An outsider can never tell an organization how to better run its business.

He illustrated the difference between efficiency and effectiveness with a story about the Empire buggy-whip manufacturing company, which at the turn of the century was the best buggy-whip manufacturer of all time. Every buggy-whip they made was engineered to specification; they rarely broke, and all grievances were promptly resolved to the customer's satisfaction. In terms of efficiency, they were among the best. The problem, he said, was that they did not have a view of the future. They were in the transportation business and did not see the coming of the horseless carriage. In ten years they were out of business because they did not know the difference between effectiveness, or doing the right things, and efficiency—doing the right things right. This may seem obvious now, but at the time it made a very powerful impression upon me.

In 1980, I joined Florida Power and Light (FPL) and was part of the original design team that initiated and implemented FPL's quality system, which eventually won the Deming Prize. The chairman at the time, Marshall MacDonald, was a genius in his own right. He knew that the same management system that had brought FPL into the 1980s would not be good enough for the 1990s and beyond. He began the odyssey, and we charted a new course that would lead into unknown waters and paths no one had traveled before, at least in America. Marshall's genius was thinking in terms of possibilities.

In 1984, Kansai Electric Company became the first electric utility in the world to win the Deming Prize, and FPL became their overseas protégé. Over the next couple of years, from 1987 through 1989, FPL began making preparations for a serious run at the Deming Prize and we began to receive firsthand counseling from JUSE, the Union of Japanese Scientists and Engineers. As the general manager of the FPL consulting and training arm, called Qualtec Quality Services, I had responsibility for translating the lessons learned for our various clients throughout corporate America. We had a pretty impressive list of clients by that time, having worked with such notables as AT&T, Eli Lily, Boeing, Chase Manhattan Bank, the IRS, New York Life, and various city and state governments, as well as such universities as Michigan, Oregon State, and Miami. To round out the picture, IBM and Xerox started using the FPL quality system as their total quality management benchmark and the Deming Prize as the ultimate challenge.

Serious contenders for the Deming Prize had to meet two essential requirements, which were basically the same requirements that a Japanese company would have to meet. The first was a promise to work with other American companies, and the second was to add something new to the total quality model. The challenge at the time was that no non-Japanese company had ever won the Deming Prize, much less an American company.

In order to comply with the first requirement and remain on good terms with the Florida Public Service Commission (FPSC), Qualtec

Quality Services was formed with a core group of five or six people. The idea was to recover the cost of sharing the lessons learned, make a reasonable profit, and not burden the Florida rate payers with the expense of doing business. The requirement was fairly easy to deal with, because FPL had developed a quality system that would prove to be transferable to every type of business and industry imaginable.

The second requirement—adding something new to the total quality model—led to uncharted waters. A portion of that something new turned out to be the Malcolm Baldrige National Quality Award, for the Deming Prize influenced the creation of the Baldrige Award in three ways. First, FPL played a behind-the-scenes role in gleaning information from Deming Prize winners, which was used to set up the initial Baldrige categories. The FPL director of quality at the time was a tall (6′6″), slender visionary who walked around with his shirttail hanging out most of the time. His name was Kent Sterett, and he reported to Bud Hunter, who was a somewhat shorter (5′6″) senior vice-president and master politician. Together they made a great team, and I affectionately dubbed them Mutt and Jeff. These two did a lot of the background work in framing the Baldrige effort at FPL. In fact, Bud Hunter has never received the recognition he deserves for his efforts and for the risks he took, which at the time were considerable.

The second influence of the Deming Prize was that it was internationally recognized as being representative of the best total quality management practices. This was an important criterion in the creation of the Baldrige Award. The final influence was that winners were obligated to share good practices and transfer the information learned to other companies.

During this developmental period, Dr. Deming visited Miami a few times to conduct his seminars. On one occasion, we informed him of FPL's intention to pursue the Deming Prize. He said that FPL was not ready. We asked him who was ready, to which he replied, "No one in America is ready!" When John Hudiburg, who succeeded Marshall MacDonald as FPL chairman, heard this remark, he was undaunted and seemed more determined then ever that FPL would win the prize.

The real test and unrest were yet to come, however. At the beginning of the summer of 1989, after returning from a steering committee meeting in Tokyo, Kent Sterett resigned and went to work for Union Pacific Railroad. He was the first casualty of the Deming

Prize, but certainly not the last, as we were soon to learn. During the final stages of the Deming Prize audit, Marshall MacDonald retired as board chairman. Jim Broadhead, who came from GTE, took his place. He was a newcomer to both FPL and the industry. The full effect of this change was not felt until the end of the summer of 1989, a couple of weeks before the audit was to be finished. Suddenly, John Hudiburg retired—the same John Hudiburg who, along with Sandy McDonnell of McDonnell-Douglas, had raised the $11 million seed money for the Malcolm Baldrige Award and who, along with Dr. Juran, had petitioned Congress to sponsor and approve the Baldrige Award.

The news caught many people, including me, by surprise. While we were still recovering from the shock, I received a phone call informing me that Mr. Broadhead wanted to see me. By this time, it was common knowledge that the new chairman was in favor of neither the Deming Prize nor the highly structured quality system that FPL had in place, and he made no bones about it. Naturally, I wondered what was next. At a meeting with his top lieutenants, the chairman asked me what the impact on Qualtec would be if the Deming Prize challenge was halted. Without hesitation, I told him that Qualtec would be out of business in a year, our clients would feel betrayed, and we would probably be sued and vilified by numerous heavy hitters throughout America who had adopted the FPL system as their model.

He then gave me some good news, which was to make my day. It seemed that a certain journalist by the name of Mary Walton wanted to do a story about FPL and the Deming Prize experience as part of a book she was writing. She had contacted our public relations department on a number of occasions, only to be turned away each time. Finally, she appealed directly to Jim Broadhead. He made it perfectly clear that while he was not in favor of seeking publicity, he would be willing to entertain the possibility of honoring her request if it would in some way benefit Qualtec's marketing efforts. He told me to see our public relations people and check out this Mary Walton. He also made it crystal clear that he would hold my feet to the fire should anything blow up in our corporate face.

After checking with Public Relations, I contacted Mary to get some of the particulars about her book project. I was intrigued—but cautious. I told her that it was a touchy situation, which she already knew, having become frustrated at this point by the rebuffs and delays. I promised her that I would see what I could do, and I called Dr. Deming to inquire about his support for Mary's project. Yes, he told me, the book had his full support, even though Mary, and not he, had select the companies. Although he would not write any of the text, he would write the foreword and would review the manuscript. It was perfect; a surefire way, I believed, to prove to everyone once and for all that the FPL journey was for real and not just smoke and mirrors and undue complexity. Better still, the story would be seen through the eyes of a trusted journalist.

I was convinced, and now all I had to do was convince our Public Relations Department to go along with the project, which turned out to be not too difficult. The plan was for Mary to come to Florida for the Deming Prize announcement, slated for the week of October 23, 1989. While not yet official, we kind of knew by the end of August that the prize was ours, barring any major unforeseen disaster. All of Mary's activities would be handled through Qualtec, and as the general manager I was to be personally responsible for the final arrangements. It was also decided that I would be Mary's personal escort, and we made plans to spend the better part of the week together. Mary agreed to run any final manuscript past me for dissemination and final review.

Everyone agreed that it was a workable plan, and we were off and running. My only concern at this point was to ensure that the whole interview process was natural and not overcontrolled. Mary and I agreed that she would be allowed to talk to anyone she wanted to interview, based upon their availability, and that included the recently retired chairman, John Hudiburg.

The day of the official announcement, we were in the general office with Bob Tallon, who was president at the time. After Bob made the announcement to a cadre of local reporters, Mary suggested that we call Dr. Deming and tell him the good news. After getting through to him via his trusty assistant, Ceil Kilian, Mary told him that she was at FPL and then gave the phone to me. After exchanging a few pleasantries, I broke the news that FPL has just become the first American company to win the Deming Prize. "What do you think?" I said, "isn't this great news!" There was a long pause and then came the comment that I will never forget: "So?" So, I said to myself—so? For a moment, I was completely stunned; I felt like I had been hit over the head with a two-by-four. Then I recovered my composure a little and mentioned all the months and years of hard work and effort by all the thousands of people to achieve this extraordinary accomplishment. I could tell that Mary was as surprised as I was, maybe even more. I think she truly believed that he would soften his position and recognize the FPL accomplishment.

I wanted desperately to move this mountain of granite of a man, whom I respected and admired so much, but I could only think of one more thing to say. I mentioned all the FPL Qualtec clients, many among America's Fortune 500, that were implementing quality systems based on the model that won the Deming Prize, which many of them believed to be the Deming method. I saw Mary smile, and I knew that I had made my point, even before Dr. Deming closed with one of his usual pearls of wisdom:

> What is really important is to decide what's next. The prize by itself is meaningless and can be destructive if not carefully dealt with. Chasing prizes is like chasing too many rabbits. You can never catch them all and you'll get worn out in the process. So challenge them with—what's next.

After we hung up, Mary and I decided to get some lunch. John Hudiburg joined us. People kept coming over and congratulating John in a very matter-of-fact way, and I sensed that Mary was becoming a little concerned. After John left, I asked Mary what was wrong. She said that it just didn't seem like a celebration—there were no balloons, no partying, no fanfare. It seemed more like business as usual. I had to laugh because Dr. Deming had just finished telling us to avoid the prizes, slogans, and fanfare. Actually, I told her that the reason for the lack of excitement was that the announcement was anti-climatic. We knew months ago that it was a done deal, and that was when the real partying occurred. The announcement was a mere formality to most employees.

Dr. Deming's point kept burning in my mind: "What's next?" We did not have to wait long to find out. Bob Tallon, president of FPL at the time, joined us later in the day. After Mary finished her interview with him, I told him about the call to Dr. Deming that morning and repeated his question: "What's next?" "Darned if I know," Bob said in his usual laid-back, casual way, which was his trademark as a great leader and executive. "We're looking into a sponsoring role in the upcoming Olympics as a possible way to keep the momentum going."

Unfortunately, the Olympics never materialized and the bloom quickly fell from the rose. In a couple of months, Bob would announce his retirement, and I would come to deeply regret his leaving because he is one of the finest gentlemen I have ever known. He also gave me a portent of things to come. Within one year, fourteen of the top eighteen executives and directors who were involved with the prize were gone. Wayne Brunetti became the chairman of MSI, an international consulting organization. Bill Hensler, who had replaced Ken Sterett at FPL as the director of quality, came to work for me as director of consulting. In fact, many "retired" or resigned and worked for Qualtec over the next near or so as executive consultants, and my office became a popular place. My Sancho Panza from the Bronx, Irwin Weinberg, also came on board, along with, of all people, John Hudiburg, who had decided to become an executive consultant.

Meanwhile, Jim Broadhead began streamlining the FPL quality system that had been the focus of the Deming Prize examination. Make no mistake, some of this was desperately needed, wanted, and welcomed by employees and management alike. I call it the "Curse of the Prize" because of the enormous toll that it took on the organization's systems, especially the social system or culture. In the Olympics, the structures that must be put in place to ensure success; or at least maximize the chances of it, become overly complex and burdensome once the medals have been passed out. Similarly, the FPL system was in desperate need of streamlining and a breath of fresh air.

Mary finished her manuscript of the FPL story, which became the opening section of her second book, *Deming Management at Work*. At FPL alone, she talked with over 200 workers and was pretty much able to pick and choose who interviewed. I made certain suggestions and provided access privileges to any of the FPL site locations within reach during the week. She covered a lot of ground and asked a lot of questions, and I think in many ways I learned as much or more from her that week as I did from the many consultants with whom I have worked. What emerged in the section of her book about FPL is a portrait so accurate, so alive, and so moving that I could not wait to get my hands on the finished book. Since early on, I had intended to use the data to convince Dr. Deming; now I was going to try to convince others as well. After all, Mary was an objective third party and had no ax to grind. She was telling it like it was, or how it looked to her at the time, and her reputation for honesty and for outspokenness was above reproach.

Unfortunately, the chairman of FPL had other plans. His famous "letter to employees" created a furor across corporate America, the likes of which had rarely been seen before. In his much publicized internal letter to FPL employees, he laid out a seven-point plan for massive streamlining and restructuring. Somehow that letter found its way into the quality department and boardroom of seemingly every organization in America. Many people used it as an excuse for the lack of performance of their own quality systems. My phone never stopped ringing off the hook. I must have spent at least thirty hours a week over the next six months making presentations and trying to explain things to countless senior executives and their direct reports, who were trying to keep motivated and focused. Everybody wanted to know what was going on at FPL, and rightly so. All of these organizations had a lot riding on their quality journeys, and if FPL, which was the beacon of light at the time, was perceived as abandoning its award-winning quality system, what were the implications for them?

In order to counteract the bad publicity and make as much lemonade as I possibly could from the lemons, I went on the road and started making speeches, in some cases as many as twelve to fifteen a month, in addition to running the consulting side of the business. Fortunately, I had great people working for me, about seventy-five or eighty at the time, many of whom did double time on the road as well. On my team were Bear Baila, the best "people-person" in the quality movement, and Bill (Gabby) Hayes, a mastermind who really understood Japanese management systems. The leader I reported to, Roger Eatman, was a Deming disciple and a visionary as well. We used Mary's book as the backdrop for our version of the FPL story, and she even joined me to do a day-long session for government employees in Denver, Colorado. I saw Dr. Deming twice during that period and he asked me about what was going on at FPL. I told him about some of the changes, and he remarked that what was really going on demonstrated a lack of constancy of purpose. He said we needed to get back to the basics and forget all this business about prizes.

Somehow we survived the year, and late in 1990 Myron Tribus and I did a session for the Midlands Area Council for Excellence (MACE), one of the community councils he helped create with the assistance of Dr. Deming. I told the story of the Deming Prize, including its history and application at FPL, and all that transpired as a result. Myron persuaded me to expand and publish the talk, which I did in the *South Carolina Business* journal, as a special issue called "In Pursuit of Total Quality." It was my swan song at FPL, and a year later I left as well. It is offered here once again just to set the record straight. For as sworn defender of the faith, I could not leave in honor until the "Curse of the Prize" had been avenged. Dr. Deming, you were right once again. The problem with prizes is that you take your eye off the customer—who is the only prize to be won and cherished, after all is said and done.

Let us always cherish the mind and the memory of W. Edwards Deming, a giant of a man among many great minds around him. A person who could truly say, "I have done my best!"

The Deming Prize*

by Frank Voehl

When the Supreme Command for the Allied Powers (SCAP) established itself in Japan after World War II, its primary goal was the dissolution of that country's military government and the establishment of a constitutional one in its place. In achieving that end, SCAP operated as an occupying power. Japan had given up its industry and eventually its food supply to support its war effort. Subsequently, there was little of post-war Japan left to occupy. The country was a shambles. Only one major city, Kyoto, had escaped wide-scale destruction; food was scarce and industry was negligible.

* Reproduced with permission from *South Carolina Business*, Vol. 11, pp. 104–110, 1991.

Frank Voehl is a Visiting Professor at Florida International University. He is President and CEO of Strategy Associates and CEO of Anro Metals Manufacturing Company. He is Series Editor of the St. Lucie Press Total Quality Series and has authored hundreds of papers, articles, and books.

Against a backdrop of devastation and military defeat, a group of Japanese scientists and engineers organized appropriately as the Union of Japanese Scientists and Engineers, or JUSE, and dedicated themselves to rebuilding their country. Reconstruction was a daunting and monumental task. With no food or immediate means of producing it, export of manufactured goods was essential. However, Japanese industry—or what was left of it—was producing inferior goods, a fact recognized worldwide. JUSE was faced with the task of drastically improving the quality of Japan's industrial output as an attractive exchange commodity for a most basic necessity—food.

Through some American engineers working with the SCAP, JUSE learned of a book containing statistical quality control techniques. These techniques, they were told, were used by U.S. industry during wartime production. The book was *The Economic Control of Quality Manufactured Products*; its author was Walter A. Shewhart.

Shewhart, a statistician with Bell Telephone Laboratories in New York, was responsible for the development of techniques that could be used to bring industrial processes under statistical control (a phrase he coined). He also believed that workers could be taught to use these techniques themselves, thus enabling them to make their own job adjustments. Shewhart developed these techniques in the 1920s and 1930s. It was during this time that another statistician came to study under him, a young man who would eventually take Shewhart's work and use it as the basis for a complete system of quality improvement That young man's name was W. Edwards Deming.

Copies of Shewhart's text were shipped from the United States to Japan. JUSE was fascinated by Shewhart's theories and their applications. Upon further study, JUSE realized that Dr. Deming had worked with Shewhart. Some JUSE members knew Dr. Deming personally, as he had worked in Japan as a statistician in 1947. They felt that perhaps he could help them in their quest to change their economy. In 1950, Dr. Deming was invited to address an assembly of engineers, plant managers, and research workers. In June of that year, he delivered the first of what would multiply into twelve seminars. Five hundred people were present at his initial address.

Although enthusiasm for his techniques was evident, Dr. Deming knew that for those techniques to work, top management would have to be present and intimately involved. He learned this from businesses in the United States that had used statistical quality control during the war effort. In the United States, middle managers and workers had been trained, and the success of the process peaked during the height of the war effort. As World War II wound down, however, statistical quality control was dropped by top management, ostensibly because of cost and the amount of time needed for training. It was deemed impractical. Dr. Deming, therefore, felt that he would need to sell Japan's top executives on the process if it were going to succeed.

The president of JUSE at that time, Dr. Ichiro Ishikawa, obliged Dr. Deming by having twenty-one presidents of Japan's largest companies join the American for dinner. These Japanese business leaders were so impressed with Deming after that July dinner that he was asked to address them again. A few of the things that impressed them were Deming's emphasis on the customer as the main part of the production line, his contention that they should not accept inferior supplies and try to fit them into existing designs, and his insight that they could demand quality supplies, redesign their products, and bring them under statistical control. Dr. Deming also predicted that if they followed his advice, they would have the world at their industrial doorstep in five years.

By August of 1950, Deming had spoken at the Tokyo Chamber of Commerce to fifty more manufacturers. By the end of the summer, he had addressed the management of most of Japan's large companies. The information he shared was taken to heart by the Japanese. Almost all Japanese companies rigorously applied statistical methods and process controls. In 1951, the Japanese created a quality award to be given in two categories: (1) to an individual for accomplishments in statistical theory and (2) to companies for accomplishments in statistical application. The award was named the Deming Prize.

In 1960, in a pamphlet written for the tenth Deming Prize awards, former JUSE Managing Director Kenichi Koyanagi wrote, "Special mention must be made of the fact that the Deming Prize was instituted with gratitude to Dr. Deming's friendship as well as in commemoration of his contributions to Japanese industry. When Dr. Deming gave his eight-day course in 1950, Japan was in the fifth year of Allied occupation. Administrative and all other affairs were under the rigid control of the Allied forces. Most of the Japanese were in a servile spirit as the vanquished, and among Allied personnel there were not a few with an air of importance. In striking contrast, Dr. Deming showed his warm cordiality to every Japanese whom he met and exchanged frank opinions with everybody...He loved Japan and the Japanese from his own heart...Herein lies why we loved and respected, and still love and respect, him."* Dr. Deming had predicted that Japanese products would be in heavy demand in five years; they were world-class in four.

Today, the Deming Prize's purpose, as stated in its guidelines, is to "award prizes to those companies which are recognized as having successfully applied CWQC (Companywide Quality Control) based on statistical quality control and which are likely to keep up with it in the future." While statistical techniques are emphasized overall, the criteria for judging are broken down into ten major categories and sixty-three subcategories. The ten major categories are: (1) policy and objectives, (2) organization and its operation, (3) education and its extension, (4) assembling and disseminating information, (5) analysis, (6) standardization, (7) control, (8) quality assurance, (9) effects, and (10) future plans. The subcategories are designed to define major areas of emphasis. They are sparsely worded, and this is done deliberately to ensure that guideline definitions can change as dictated by time.

Not all Japanese companies are concerned with capturing the Deming Prize. Prize applications sometimes run up to one thousand pages, and the whole process, including site visits, sometimes takes years. Some Japanese companies are simply unwilling to invest the amount of time or money needed to make a legitimate run at the Deming. It is regarded worldwide, however, as a standard in quality. Some past winners include Sanyo Electric Works, Ltd., Fuji Xerox Co., Ltd., and Toyota Auto Body Co., Ltd.

It is ironic that the same methods that were in part responsible for maximum output by the United States in wartime production were abandoned once the war was over, because they were considered too costly and time-consuming. It would take nearly thirty years for U.S. industry to feel the full repercussions of that mistake. In those years that the United States enjoyed a captive world market, the Japanese were hard at work improving and attempting to perfect quality techniques. Around 1980, some of the leading U.S. companies awoke to a desperate need. Foreign products were competing favorably in nearly every market, and "Made in America" was not

* Kenichi Koyanagi, *The Deming Prize,* Union of Japanese Scientists and Engineers, 1960.

quite the assurance of quality that it used to be. Many U.S. companies came to realize that money spent on quality was an investment, not just an expenditure; in short, it was a strategic move to ensure survival.

In many ways, Florida Power and Light has been just such a company. As a utility, it has never been threatened directly by foreign competition, but in 1981 domestic problems provided ample impetus for change. Faced that year with inflation, rising fuel oil prices, and the prospect of large capital expenditures to meet future demand, the company realized the need for a plan to implement quality extensively. Such a plan would have to equip the company with a resiliency to cope with changing and often hostile circumstances. In the late 1970s, quality assurance programs for nuclear power plant construction had proven successful in terms of cost containment and timing, quality control, and public safety. Other sections of the company were encouraged to follow suit, and their individual successes led, in 1981, to a personal commitment to the concept of total quality by then former Florida Power and Light Chairman Marshall MacDonald.

Commitment by top management was the first step in Florida Power and Light's quality process. In a move that has come to be somewhat mandatory for leading U.S. companies, Florida Power and Light officials then visited Japan to experience firsthand the benefits of stringent application of quality techniques. They toured Kansai Electric Power Co., a company that would, in 1984, win the Deming Prize. A quality improvement plan was initiated; experts such as Dr. Deming were consulted. These events led to the formulation of Florida Power and Light's Quality Improvement Program (QIP). This program's mission is "to become, over the next decade, the best managed electric utility in the U.S., and an excellent company overall, and be recognized as such."

In keeping with this plan, FPL has, since 1983, spent more than $75 million on computers. This extensive information network has improved response time to customer complaints, power plant repairs, and installations. The company is becoming a role model in the use of technology to aid quality. For example, its Generation Equipment Management System (GEMS) tracks equipment at Florida Power and Light's thirteen oil-fired fuel plants. When a generator goes down, the system automatically budgets the repair, orders parts, and calls in a work crew. Also, through the historical database it has

Comparison of Baldrige Award and Deming Prize

	Baldrige Award	Deming Prize
Origin	• Created and first awarded in 1987 after President Reagan signed U.S. Quality Improvement Act into law	• Created in 1950 by Union of Japanese Scientists and Engineers (JUSE); first awarded in 1951
Guidelines	• Seven categories containing approximately sixty sub-items • Guidelines intended as blueprint for implementing Total Quality Management (TQM)	• One-page copy of general, loosely worded guidelines furnished by JUSE upon request • Criteria based largely on judgment of JUSE counselors
Application process	• Sixty-five pages of questions • Description of Quality Improvement practices (DQIP) not required	• Six pages of questions • Description of Quality Improvement practices (DQIP) required
Written examination format	• Questions presented in structured, multiple choice and/or completion formats	• JUSE counselors review and critique applicant several times over twelve-month period • Questions presented in thesis style
On-site examination	• Conducted over three-day period by team of four to five persons • Team's primary objective is to clarify unclear aspects of application	• Conducted during two or more weeks with forty examiner days at any site • Examiners may query any employee and ask to verify comments with data

accumulated on equipment performance, GEMS can predict mechanical breakdowns and warn maintenance crews of potential malfunctions before they occur. Implemented in 1986, this system, in just over two years, saved the company over $10 million and cut fossil-fuel plant downtime from 14 percent to 8 percent.

In July 1988, under the direction of then former Chairman John J. Hudiburg and President Robert Tallon, Florida Power and Light opened eyes everywhere by applying for Japan's Deming Prize. This was remarkable for two reasons: (1) Japanese product and process quality was (and continues to be) world-class in every respect and (2) no company outside of Japan had ever won (or even applied for) the Deming Prize. As unheard of as it was, however, FPL's unwavering commitment to quality paid off, and in 1989 the company was awarded the Deming Prize. This commitment has also paid off in more concrete terms—customer complaints and fossil-fuel plant downtime are both at an all-time low.

In addition to providing leadership in Florida Power and Light's quest for the Deming Prize, former Chairman Hudiburg has been to Japan and has been instrumental in the U.S. quality movement. He testified before the Congressional Subcommittee on Science, Research, and Technology on the need for a national quality award. He stated that the United States needed a monument to quality if it was to become a leader again in the world marketplace. In his testimony, Hudiburg said that "the pride of 'Made in America' is being rediscovered in a revival of quality—quality products as well as service. I believe that it's time to encourage, recognize, and reward that effort." As a result, U.S. business now has its own award to aspire to, the Malcolm Baldrige National Quality Award. Recipients of this honor are considered on a par with the world's best in their respective fields. Hudiburg was also responsible, along with Sanford McDonnell, for creating the Baldrige Foundation, a foundation of private business and industry that donated millions of dollars to supplement the award program. This management commitment continues with the current FPL Group chairman serving as a trustee for the Baldrige Award.

It should be noted that in 1981, even though strong economic influences were present, Florida Power and Light did not have to embrace the all-encompassing quality program that it did. As an electric utility, the company was already heavily regulated, and industry advances in technology are usually few and far between. It is a tribute to the people and management of Florida Power and Light that they opted to travel the harder road, because in doing so they became an inspiration for many U.S. companies and were responsible for many industry advances of their own.

It seems odd, somehow, that an American export has found its way home and after nearly forty years is being utilized with zeal. The pride that was once representative of American-made products is steadily returning. As the saying goes, "Physician heal thyself." Americans created the original industrial quality processes, Americans exported it to help a country survive, and Americans have recently found themselves in a survival situation, businesswise. Often we are slow to accept, but we are always quick to learn. Quality is contagious, and because U.S. enthusiasm is unique, it is only a short matter of time before we take our place once again at the top of the world marketplace, with our eyes set on the future. And the phrase "Made in the USA" will once again be the standard of excellence throughout the world.

There are many hazards to the use of common sense. Common sense cannot be measured. You have to be able to define and measure what is significant.

W. Edwards Deming
The Keys to Excellence

Chapter Thirteen

REMOVE BARRIERS TO PRIDE OF WORKMANSHIP

"Remove barriers that rob the hourly worker of his right to pride of workmanship. The responsibility of supervisors must be changed from sheer numbers to quality. Remove barriers that rob people in management and in engineering of their right to pride of workmanship. This means, inter alia, *abolishment of the annual merit rating as well as Management by Objectives, management by the numbers."*

Dr. Deming believed that workers want to be proud of their work and resent it when they have to perform shoddy work in order to get the job done. He tells the story of poor old Joe, who was so beaten down that he no longer cared—he had been robbed of his motivation through management by fear techniques. Deming taught that workers like Joe have a right to be proud of their work. Once the obstacles are removed, the workers begin to blossom. Removing the barriers, however, is the job of management, which is a recurring theme we hear Dr. Deming emphasize over and over again.

Notice how Dr. Deming carefully lays out the reason for the action before identifying the three villains of quality. He begins by clearly identifying the call to action: remove barriers. Rather than form a team, develop a plan, or sing a song about improvement, he begins with a simple direct charge: remove barriers. Which barriers? Remove those that inhibit pride of workmanship. How? Shift the focus from sheer numbers to quality.

Dr. Deming does not stop there. He goes on to involve managers as well as engineers in this crusade by doing away with the dreaded villains management by objectives, management by the numbers, and performance appraisals. For a while, he had little use for quality circles either, even though he saw their power in transforming Japan. He viewed these tools as gimmicks and quick-fix solutions. They give supervisors and managers a shield to hide behind as they duck their responsibilities. They might be useful for a while as a smokescreen so that management can point to something being done, Dr. Deming says. However, the programs eventually begin to fade away because management never invests the workers with true authority. In other words, they lack empowerment.

When management stops listening to the workers' decisions and recommendations, employees become even more disillusioned. To illustrate this point to a client's management team, Dr. Deming would often begin with a "workers only session," during which members of management could hide behind a curtain in the projection room and tape the session for future viewing with the quality council. It was at these sessions that Dr. Deming was really at his best. He was a master instructor who was exceptionally skilled at using dialogue to draw out the workers. And what dialogue it was—skill channeling of conversation toward an objective, almost like lines from a play. A hundred years from now, they will be called the Dialogues of Deming. He was influential in reviving the essence of dialogue, namely, the informal exchange of views and communication between parties in a system. FDR had his fireside chats and Dr. Deming had his dialogues.

He disliked the concept of performance appraisals so much that they found a special niche as one of his Seven Deadly Diseases. He called them one of the *inter alia*, or among other things, and he illustrated the problem in the form of a control chart. At his talk on the occasion of the thirty-fifth anniversary of the Deming Prize in

Japan, he showed how rating people in a group by any numerical system, however perfect, will divide the group into three categories. Group A includes those who score outside the control limits on the negative side. Group B includes those who score outside the control limits on the positive side. Those who are between the limits are in Group C. People in Groups A and B, if there are any, he says, must be dealt with individually and swiftly because they need prompt attention.

"Stop! Go no further," he would say. "If you try to rank the individuals within the control limits, which constitute the majority, you destroy the morale and well being of the system." These are strong words, but he meant every one of them. His statistical genius and training had taught him that it is costly and non-productive to try to determine why one data point is higher or lower than another. We can never be sure, because the answers are buried deep. The truly wise know that no matter how deep they dig, the answer is buried deeper still, and the truly simple know that there is no use in digging at all. Differences within the levels come from the system itself, not from the people involved. He believed that everyone in Group C should receive the same increase in pay or the same bonus, although not necessarily the same salary. The job of the leader of the future is to shrink the control limits, to get less and less variation in the system or process, without squashing the individual or individual creativity.

My friend Lou Schultz is a master at teaching these skills to managers, and his company, PMI, located in Minneapolis, has been teaching the Deming philosophy all around the world. Lou is a mixture of the old and the new, and I am certain that you will enjoy his story.

The Changing of a Career: How Dr. W. Edwards Deming Changed My Life

by Louis E. Schultz

Dr. W. Edwards Deming was fond of closing his four-day seminars with a simple statement, "I leave you with five words: I have done my best." That he did! He believed in people and believed that people did their best in spite of faulty processes and systems that put barriers in their way. He impacted and changed many thousands of lives, including my own.

On June 24, 1980, I had my first view of Dr. W. Edwards Deming. Little did I know this chance occurrence would impact the rest of my life. On that summer evening, I had returned home from coaching a Little League baseball game and sat down in the kitchen to perform the paperwork that was required. The television set was on in the kitchen, as it usually is in our house, and I started watching, out of the corner of my eye, an interesting program entitled "If Japan Can...Why Can't We?" As everyone probably knows by this time, it was all about what a man named Dr. W. Edwards Deming had accomplished in Japan by turning their economy around and about what he was now doing in the United States. It talked about a client, Nashua Corporation in Nashua, New Hampshire, that Dr. Deming and another consultant, Charlie Bicking, were working with to improve the company's performance. The main theme of the program was how quality had been turned around in Japan, from very poor quality to a nation that was sweeping the world with excellent products. Before long, I put my pencil down and watched intently, mesmerized by the potential this had for the United States.

Louis E. Schultz is President and CEO of Process Management International (PMI), a Minneapolis-based TQM consulting organization. He has been a Deming disciple since the early 1980s. While in a management position at Control Data, he worked closely with Dr. Charles Bicking, with whom he entered the realm of total quality.

The next day, I told my boss, Larry Jodsas, who was the operations vice-president for Peripheral Products Company of Control Data Corporation, about what I had seen, and he said, "Schultz, if it's that good, get a copy of it and show it to Tom Kamp [the president of Peripheral Products Company]." I asked my secretary to call NBC and find out how we could order a copy of the tape. Meanwhile, I was talking to a friend about the program, and we concluded that I really needed more information about this Dr. Deming. I remembered that the program had said that he lived in Washington, D.C., so I called the information operator, got a number for a W. Edwards Deming, and called with the intent of having his secretary send me some literature. To my surprise, he answered the phone himself, and I did not know what to say. I had never in my life called anyone whom I had seen on television! He was a very nice gentleman. We talked for some time, and he did, in fact, personally send me some literature.

I attended his seminar in Washington, D.C. in September 1980 and was not disappointed. At the end of the first day, I telephoned Skip Akerlund, the corporate vice-president of quality at Control Data, and told him that I thought we were onto something really worthwhile. Dr. Deming's notes were in a plastic comb binder that was about an inch thick. I realized it was unrealistic to think that management would read this much material, so each evening when I returned home I tried to summarize the key points. I ended up with a twelve-page trip report which I distributed upon my return.

I invited Dr. Deming and the great statistician Charlie Bicking to meet with our management in December 1980. Dr. Deming also did a private four-day seminar for us in May 1981, which involved about 380 Control Data managers. He noted that many of the top executives with whom he had met in December were not present, and I started to worry that he would not finish his seminar. Tom Kamp eventually came over and talked to the attendees, which was not at all what Dr. Deming wanted. He wanted our executives to listen, not to talk. He wanted them to listen to their workers, who wanted to tell them about all the barriers to workmanship in their areas, but everybody is too busy to listen, too busy to pay attention, too busy to care.

We continued to work with Charlie Bicking as we engaged in a consulting contract, but Dr. Deming continued to be upset by the lack of executive involvement. Tom formed a task force to develop the quality program for Control Data, and I was named to the task force. About six months later, Larry Jodsas agreed to have me chair the effort, and hence a career change was in the wind for me. I attended an optional seminar offered at Control Data entitled "Work/ Life" and did some visioning exercises, one of which was to envision what we would really like to be doing in our careers. When I wrote down what I envisioned, it had nothing to do with working for a large company. In fact, I was working with a small group of three or four people, doing something worthwhile in a stress-free, barrier-free environment and really enjoying our activity and having pride in our work—also known as heaven.

As our work at Control Data continued to grow and gain notice, I received more and more attention, and people outside the company started inviting me to speak. It started to get embarrassing, taking vacation time to go to places such as Puerto Rico to speak about quality, and my relationship with Control Data management became strained. At this point, I started to entertain thoughts of starting my own company. It still amazes me how the chance viewing of a television program about Dr. Deming could change my life so dramatically.

When I founded Process Management International (PMI) in 1984, we sent our business plan to some people we respected and sincerely asked for their opinion and advice. To our dismay, the response from Dr. Deming was not as positive as we had hoped. He simply stated that we would be out of business in two years, but if we insisted on going ahead, he would invite me to a seminar in San Diego in January 1984 and he would take care of the fee. I did not know why he wanted me to come back to a seminar I had attended twice previously, but I certainly respected his input and did not want to go against his wishes. During my last week of work at Control Data, I took a week of vacation and attended Dr. Deming's seminar in San Diego. It was there that I began to understand what he was talking about. We were focusing only on Statistical Process Control when what was required was a transformation of the management culture of the organization. That culture change is what we would have to work on as well.

I have always been grateful to Dr. Deming for insisting that I come to his seminar one more time, because I learned so much that third time, more than I had learned the first two times. I have now been to Dr. Deming's seminars twenty-three times, and I learn more each time. There was so much depth to what he had to say that it was impossible, in my opinion, to learn it all in one sitting. Dr. Deming believed in continual learning. He listened to and learned from those around him throughout his life and never ceased to contribute knowledge in return. I found I could never relax and let my brain idle around him. Whether in a large seminar or privately with just the two of us in a car, he always forced me to think. I could never thank him enough for the impact he has had on my life, both professionally and personally.

Over the years, Deming's philosophy—including the content of the Fourteen Points, has changed, largely due to his devotion to constant, never-ending learning. For instance, in 1985, Point 14 read: "Clearly define top management's permanent commitment to quality and productivity and its obligation to implement all these principles." Today, it reads: "Involve everybody in the company to work to accomplish the transformation. The transformation is everybody's job." Another example is: "Focus supervision on helping people do a better job. Ensure that immediate action is taken on reports of defects, maintenance requirements, poor tools, inadequate operating definitions, or other conditions detrimental to quality." Today, it says: "Institute leadership. The aim of leadership should be to help people and machines and gadgets do a better job. Leadership of management is in need of overhaul, as well as leadership of production workers."

The wording of other points has changed as well, but the underlying philosophy has not: Management has the responsibility for the overall system. Managers must stop blaming workers for problems inherent in the system. They must set about fixing the system so the workers can do their jobs better. They should also cease their preoccupation with this month, this quarter, or even this year and start thinking about where their company will be in five years. Only with a strong vision of the future, solid information derived from statistical methods, and an understanding of human psychology will a company be able to compete in the rapidly changing world market. While other quality experts founded institutes or consultancies to carry on their teachings, Dr. Deming chose a less formal route, anointing a cadre of consultants and academicians as disciples. Consultants, mostly statisticians, started their own companies to disseminate his teachings. In several communities, people whose lives changed when they learned his theories formed non-profit clubs and other organizations to help each other learn more. Dr. Deming once told me that the legacy he wanted to leave was *systems thinking* and *win–win*. So simple and yet so powerful! By practicing it, he left a far greater following than if he had founded a single organization.

Dr. Deming conducted his famous four-day seminars well into his nineties, beginning them with, "Why are we here? To learn and have fun." Much of his teaching, as outlined earlier, concerns understanding psychology, variation, systems, and theory of knowledge. These headings, however, are only that—the tips of icebergs which must be viewed beneath the surface to be truly comprehended. "If you have not produced the data, you cannot understand (or use) it," Dr. Deming told his audiences. "You need to understand the production of the data."

His seminars have made a powerful impact on those who have attended. This is partially because Dr. Deming supports his behavioral theories with scientific proof, such as pointing to trade restrictions as tampering with a system and making matters worse. His seminars, in his words, were "education, not training. You will not pick up skills that will allow your fingers to do something differently on Monday morning, but you will pick up things that will haunt you the rest of your life."

Dr. Deming provoked people to think, whether in a classroom or on a drive across town. At New York University, where he taught for many years, he often liked to ask his students what other courses they were taking that would help them have a positive impact on the U.S. balance of trade. "Looking at their faces, you can see them getting mad," he said, "as they realized their other classes really do not contribute to reducing the balance of trade." "Good intentions are not enough," he continued. "They must be accompanied by profound knowledge. Quality for international trade must be good enough to command a market."

Dr. Deming's guiding star, at least one of them, was his lifelong commitment to learning and teaching. Acceptance of his teachings has been slow in the United States, but those who have adopted his philosophy and methods—from small companies to entire nations—have found themselves able to lower costs, produce more, and gain greater profits, even in markets believed to have been lost forever. This man had an effect on many lives. A couple of years ago, a magazine listed him among the top ten men who had impacted the world the most throughout history. But life goes on. The best tribute we can offer to him now is to make sure his work and teachings are carried on, so that someday people can say about us what we have said about him: we made an impact upon the world around us and we have done our best.

Later on with top management I put this question to them: why is it that with 80 percent of your capital equipment in gauges, instruments, and computers printing piles of (outputs)—and with 55 percent of your manhours going for inspection—why is it that no one except the production workers knew about the warped plates? You are (finally) concerned because your best customer is looking around for a supplier for lower prices and better quality. You may lose a customer. You can't blame him! Your prices are too high because of waste of human effort gone bad (rework, inspection) and for the huge investment in equipment for inspection and storage of useless information.

W. Edwards Deming
Out of the Crisis

Chapter Fourteen

INVOLVE EVERYONE IN THE TRANSFORMATION

"Put everybody in the company to work to accomplish the transformation. The transformation is everybody's job."

To accomplish the transformation, Dr. Deming says that management must organize as a team. Every employee must then acquire a precise, very specific idea of how to improve quality, starting with the management team or council. But how to begin? Start with the Shewhart Cycle, also known as the Deming Wheel—the wheel of continuous improvement. In America, it is often referred to as the PDCA cycle, for Plan, Do, Check, and Act. Starting in 1950, Dr. Deming used this model as the central theme for his twice-a-year conferences that he gave to engineers and managers.

The PDCA cycle is an action-oriented approach designed to involve everyone in the transformation. It is intended to be coupled with a never-ending focus on customer satisfaction—both internal and external. One of the things that I will always remember about Dr. Deming is his uncanny ability to be at the forefront of voicing new theories, as shown by his early use of the term *internal customer*. As early as 1975, he taught that each of us has customers, even if we do not interact with the external (or ultimate) customer. While this may not seem like such a big deal today, in 1975 it was positively revolutionary thinking. Along the same lines, he was an early advocate of the concept of *internal suppliers*.

Often management will require guidance from an experienced consultant, as in our private lives we may turn to our priest, minister, rabbi, or other earthly intermediary. Dr. Deming stressed that the ultimate responsibility for promoting quality improvement on a continuous basis cannot be delegated. He taught that the consultant cannot become a substitute for senior management in carrying out the Fourteen Points and eliminating the Seven Deadly Diseases. "No substitutes," Dr. Deming would say. Instead, management must adopt an evangelical zeal for studying and implementing the new philosophy, which in his mind was like the "new religion." To him, quality was like religion—not something you do once in a while and then go about your business the rest of the week. It was a way of life, a permanent and unending change of heart, a religious conversion.

If you have ever visited an award-winning quality company that practices excellence, you will know what I mean. The people you meet, from top management to the people in the mailroom, all have this evangelical look about them. They exude quality in their walk and in their talk. When you visit Milliken, IBM Rochester, Zytec, AT&T, and all the others, you come away shaking your head and marveling at their zeal and enthusiasm. They all seem to speak the same language; in other words, they are all involved in the transformation. Dr. Deming would often say, "This whole movement may be instituted and carried out by middle management, speaking with one voice." Why is middle management the point of focus? They are the hardest to motivate and often have the most to lose. They must become like the church choir, singing the songs of change. They must become disciples if the organization is to change, much less transform. They must be converted if the people are to believe.

Finally, Dr. Deming saw the organization as a team, where everyone has the opportunity to take part. He said that the aim of the team is to improve the input and the output of any stage. In *Out of the Crisis* (p. 90), he put it this way:

> A team may well be composed of people from different staff areas. A team has a customer. Everyone on a team has a chance to contribute ideas, plans, and figures; but anyone may expect to find some of his best ideas submerged by consensus of the team.

> He may have a chance on the later time around the cycle [for] a good team has a social memory. At successive sessions, people may later tear up what they did in the previous session and make a fresh start with clearer ideas. This is a sign of advancement.

In Dr. Deming's world, his church was a very important part of the team he was on. His religion was a very strong part of the fabric of his life. He composed hymns, sang in the choir, said his prayers, and followed the Golden Rule as best a man could. He believed in the universal brotherhood of mankind and treated each person he met with respect. He also believed in the universal fallibility of systems, but he balanced it all with a love for the people who created them. Even when the situation called for harsh measures, he always tried to act with a higher purpose in mind. He was a heavyweight among his peers, but heavyweights die as well. Although he has gone to his greater reward, the reward is also ours for having known him and been touched by him, if only for a moment in time.

A Requiem Offered for the Repose of the Soul of

William Edwards Deming

(14 October 1900–20 December 1993)

Christus resurgens ex mortuis, iam non moritur:
mors illi ultra non dominabitur. [Romans 6.9]

Wednesday, 29 December 1993

Saint Paul's Parish
Washington, D.C.

Requiem for a Heavyweight: Homily Delivered at the Requiem for W. Edwards Deming

by the Reverend Father August W. Peters, Jr.

On behalf of Father Martin, Saint Paul's Parish, and the family of Dr. Deming, I welcome you to his beloved church today and to this special service for him. I thank all of you for coming and for braving the elements to do so. It is a particular honor for us, and it is extremely gratifying, to have here today numerous dignitaries and official representatives of governments and corporations and so many friends and associates of Dr. Deming, not only from this city, but from across the nation and from other parts of the world. It is a fitting tribute to him. And so, again, we are grateful that you came, and we hope that you were not terribly inconvenienced by the bad weather or our space limitations, two things over which we had very little control at this time.

We are all the poorer for having lost a giant of a man, Dr. W. Edwards Deming. We are at the same time all the richer for having known him and having benefited from his knowledge, his sound teaching, his wise counsel, his kindness and generosity, and what I am bound to call the "quality life" which he led.

I cannot tell you how honored and humbled I was to be asked by Dr. Deming's family to be the preacher at his requiem. I feel privileged to do this for I have great admiration and affection for the man.

My mind ran immediately to a passage from the forty-fourth chapter of the book of Sirach, or Ecclesiasticus, which is appointed to be read at the Eucharist on All Saints' Day. As I read to you now verses one to eight of that lesson, I think you will see why I at once thought of it in connection with Dr. Deming:

Let us now praise famous men, and our fathers in their generations.
The Lord apportioned to them great glory, his majesty from the beginning.

There were those who ruled in their kingdoms,
and were men renowned for their power,
giving counsel by their understanding, and proclaiming prophecies;
leaders of the people in their deliberations
and in understanding of learning for the people, wise in their words of instruction;
those who composed musical tunes, and set forth verses in writing;
rich men furnished with resources, living peaceably in their habitations—
all these were honored in their generations, and were the glory of their times.
There are some of them who have left a name, so that men declare their praise.

Edwards Deming is a *famous man*—without doubt one of the most eminent figures, nationally and internationally, in the latter part of this century, and he is surely the most illustrious member of Saint Paul's Parish. Dr. Deming has been to many, first, a *father*—to his own daughters, to whom he was devoted, and to the multitudes of students who have sat at his feet and absorbed his doctrine. He has even been called "the father of the third wave of the Industrial Revolution." He has *given counsel by (his) understanding,* which was immense, and has been regarded by many as a *prophet* who was *wise in the words of instruction.* (I have recently been reading his books and can testify that this is true.) And so Dr. Deming is rightly regarded as *a leader of the people.* At the same time he was a man who *composed musical tunes*—we are singing some of them today for his service!—and one who *set forth verses in writing.*

Dr. Deming, as we all know, was eminently successful in his professional career as a scholar, educator, lecturer, author, and consultant. And he did it all with conviction, purpose, zeal, and moral authority. The London *Daily Telegraph* of December 24 in its article on Dr. Deming referred to "the fruits of Deming's **evangelism**" (which are found in the efficient organization of Japanese factories around the world). It also spoke of Dr. Deming pursuing "a relentless **mission to convert** ailing [American] industries to his way of thinking." It was, as he himself said, a matter of "trying to keep America from committing suicide"—certainly a worthwhile mission!

Now I know practically nothing of economics and the science of statistics, and even less of the principles of business management—fields in which he was an acknowledged expert. But I do grasp the significance of Dr. Deming's commitment to "total quality" at every level. God bless him for that important principle which he never ceased to hammer home and apply. His famous seminars obviously have made great sense to a great many people and have led to positive changes in industry—which is a thing that is good for everyone. Quite rightly, he has been *honored in his generation* and could be considered among those who are called in Sirach *the glory of their times.* Edwards Deming is a person who *has left a name, so* that *men declare his praise.* How proud we are of him today and how thankful for his time among us and for his enormous contribution to the betterment of our lives.

Having said all that, I must point out that my purpose here today is not to deliver a eulogy to Dr. Deming, recounting all the significant events and achievements of his long and distinguished life and career and extolling his many virtues. First of all, there wouldn't be time! (Some of you have planes to catch.) Secondly, it would be presumptuous of me, one of his pastors but not one of his peers, to attempt so formidable a task. How could I begin to detail with accuracy, insight, and perspective such a life and to "sum it all up" for you, so to speak? And thirdly, Dr. Deming's family specifically directed me not to do it. I am being an obedient servant of God's people. So I now must practice some restraint.

The reason why we gathered here at Saint Paul's, in the snow and ice, is because this is Dr. Deming's own parish church and this is a religious service—specifically a requiem mass, which is the service we have for every member of our Church at the time of their departure out of this world. It is offered to the glory of God, in thanksgiving for the life of the deceased Christian and for the repose of his soul. It is, further, an act of Christian faith and charity and an occasion of prayer for the consolation of the family and friends of the departed. And it is the primary purpose of the preacher on such an occasion to declare the word of God in a way that will be helpful to God's people.

We turn then to the Scriptures for a lesson from the Second Letter of Saint Paul to the Corinthians. There we hear the Apostle say to the Church:

> *We do not lose heart. Though our outer nature is wasting away, our inner nature is being renewed every day. For this slight momentary affliction is preparing us for an eternal weight of glory beyond all comparison, because we look not to the things that are seen but to the things that are unseen; for the things that are seen are transient, but the things that are unseen are eternal.*

Dear friends, the Holy Bible teaches us in many places that God has made us to be "like" Himself. That means, first of all, that we are to manifest His character and live "godly" lives in this world. We are to "be holy" even as He is holy. We are to "love one another" even as Christ has loved us. We are to do the work God directs us to do in righteousness and truth, as **His servants,** and for the common good. This is God's word to us.

I think that Dr. Deming's commitment to "total quality" in all things and at all levels had its roots and origin in his deep religious faith. I think that his own "quality life" was a concrete expression of his inner belief in God who is at work in us to transform us and to make his whole creation new and glorious. I think that Dr. Deming's character and exemplary life and his mission on earth to transform human systems and to strive for quality at all levels is the outcome of his desire, in the proper Biblical understanding of it, to be "like God"—and that means a life **submitted to God** and **oriented** toward **His** truth and the accomplishment of **His** work. There is something of the gloriousness of God in it when men, like Edwards Deming, become *the glory of their times.* Thanks be to God for that.

But, as Saint Paul shows us, in this mortal life we are bound to age and to suffer affliction. *Our outer nature is wasting away.* Sickness will overtake us and death will eventually take us out of this world.

I came to know Dr. Deming very well through his illnesses. I was one of the priests who would visit him often and pray with him and bring him the sacraments of the Church. He was most reverent when we prayed and devout in signing himself with the cross and in receiving Holy Communion. Sometimes he would pose theological questions and I would try to answer them (not always to his satisfaction!). Always he received me with the utmost courtesy and good humor and he was literally profuse in his expressions of gratitude that I had come to him. ("Thank you very much for coming; thank you very much for coming; thank you very much for coming...") It really was remarkable, I thought, because often people who are very ill will indulge in self-pity and the placing of blame for their misery. Sometimes they even reject the Church and its ministry. But Ed Deming did none of that. Which is not to make him out to be a saint far above all the rest of us, but it does suggest an uncommon awareness of what Saint Paul calls *the things that are unseen* which *are eternal.* And **that makes a difference** in a person's life.

Dr. Deming was **in every way** a "quality" man. I think he knew something about being the servant of God in truth and for the common good. Which, I believe, gave him the commitment and the fortitude not only to make such an enormous contribution to the world but to be a faithful, generous, and loving man, **even through suffering**. It also **prepared him** for **the life to come**. He *looked to the things that are unseen;* he apprehended that which is *eternal.* Which is to say that he was able to glimpse something of the glorious God who is drawing us to Himself and who is *preparing us* ***for an eternal weight of glory beyond all comparison***—and that is **TOTAL QUALITY**, THE RESURRECTION LIFE.

And so Saint Paul writes:

> *For we know that if the earthly tent we live in is destroyed, we have a building from God, a house not made with hands, eternal in the heavens. Here indeed we groan, and long to put on our heavenly dwelling, so that by putting it on we may not be found naked. For while we are still in this tent, we sigh with anxiety; not that we would be unclothed, but that we would be further clothed, so that what is mortal may be swallowed up by life.*

It never ceases to amaze me that whenever I came to visit Dr. Deming to bring him the sacraments in his illness and when he wasn't up to coming to church, I was not permitted to enter into his presence unless and until he was fully dressed in shirt, tie, and business suit! (The only exception was when he was confined to bed in Sibley Hospital and couldn't get dressed up; but he was never comfortable about that.) He was a gentleman of the old school, I guess you would say, and I gather (from one of his memoirs published in Ceil Kilian's book about him) that it goes back to the way his mother required him to dress every day when he was a young boy. But then I saw in that a symbol, a symbol of what Saint

Paul is saying. The life of the world to come is not a matter of shedding our old body of mortality and just drifting off into the great beyond like a vapor or something. This would be a denial of our essential nature as human beings. Instead, it's a matter of **being clothed with immortality—"getting dressed up" in the glory of Christ—"putting on" the new creation** which comes from God, so that *what is mortal may be swallowed up by life.* This is what it is to be ***A COMPLETED* HUMAN BEING: "TOTAL QUALITY" HUMANITY, in the image of God.**

We can all be thankful that Dr. Deming was such a celebrated and decorated human being while he lived among us. I know how much he valued the award he received in 1960 from the Emperor of Japan, the Second Order Medal of the Sacred Treasure. It's the first thing he wanted to tell me about and show me a representation of when I entered his house for the first time several years ago. And it pleased him to a great degree to have the Deming Prize established in his name to recognize Japanese manufacturers who made significant advances in the quality of their products. Over the years his superhuman (and it was that) lecturing schedule brought him considerable prestige and material reward. Yet he never became overly impressed with his position and wealth, nor did he allow himself to become secularized. He was no slave to mammon. How succinctly the London *Daily Telegraph* article put this in its concluding sentence: "He [Deming] lived modestly and gave much of his income to his local church and to medical charities." The reward of eternal life was more important to Edwards Deming than the rewards of this world. The glories of God's heavenly kingdom outshone for him the glitter of earthly success. He never forgot **who he was** and **Whose he is**.

W. EDWARDS DEMING:
child of God,
friend to man,
faithful brother in Christ

We shall not forget him. In love and gratitude we commend him to God's eternal keeping and glory. May he shine with the radiance of Christ and rejoice in the fellowship of all the Saints, now and forever, and unto ages of ages. *Amen.*

Dr. Deming would have wanted you to know the names of the participants in his requiem:

The Reverend Doctor Richard Cornish Martin
Celebrant

The Reverend August W. Peters
Assisting and Homilist

The Reverend Gregory G. Harrigle
Assisting and Old Testament Lector

The Reverend E. Perren Hayes
Assisting and New Testament Lector

Jeffrey Smith
Organist and Choir Director

The Choir of Saint Paul's Parish

David Parker
Baritone soloist

Amanda Guilford
Soprano soloist

Lowell I. Miller
Master of Ceremonies

Members of the Saint Paul's Parish Guild of Acolytes

Herbert B. Thompson
Head Usher

The Ushers of Saint Paul's Parish

Victory

Words: Latin, 1965; *tr.* Francis Pott (1832–1909), alt.
Music: Giovanni Pierluigi da Palestrina (1525–1594); *adapt.* and *arr.* William Henry Monk (1823–1889).

EPILOGUE

Dr. Deming would end his seminars and workshops with the words, "I have done my best." This did not mean that he had covered all the material in his Fourteen Points, but that he had totally immersed himself. Each week, over and over, he would hammer away and push himself beyond all human endurance to reach out one more time. He gave of himself all too often and sometimes must have felt that he gave too much to the world and not enough to his own family. All great men must at times feel this tug-of-war, as only one who is at the fulcrum can. Dr. Deming knew, as the curtain was drawing to a close, perhaps for the last time, that his words were not hollow. They had conviction and a power to them that is difficult to describe.

I was recalling this closing to Dave Schwinn as we were discussing his and Carole's contribution to this book. Dave reminded me that it was not the complete ending, because Dr. Deming would also say, "And don't forget the Seven Deadly Diseases." It was as close to hearing a voice from the grave as I care to encounter, for when Dave said it, it was almost like Dr. Deming taking me to task for being so careless as to omit the Seven Deadly Diseases.

So here they are, sir, along with this book. Consider this a final tribute to celebrate your life.

The Seven Deadly Diseases

by Frank Voehl,
as inspired by Mary Walton

Dr. Deming was a student of both history and human nature. He saw in the history of the quality movement the tendency of management to exploit the workers. He saw firsthand at the AT&T Hawthorne Lab in Chicago the same slaughterhouse mentality that he saw in the Chicago stockyards, only it was more civilized and mechanized in the stockyards, because their management had learned how to eliminate disease, while the American management system had become the disease.

He once told the U.S. Agency for International Development to "export anything to a friendly country, except American management." Management in the United States suffers from deeply entrenched diseases that are potentially deadly unless they are eliminated, eradicated, and corrected quickly.

To overcome these diseases, Dr. Deming demanded no less than a complete shake-up of the Western style of management. Furthermore, these Seven Deadly Diseases overlap and connect as part of an evil system of their own. When they are coupled with other factors and obstacles, an organization is often crippled and brought to its knees, eventually to wither and die. It may continue on for a while longer, he would say, but it eventually atrophies in a progressive manner.

Inconsistency of Purpose

If constancy of purpose is essential for a company to stay in business, then the lack of it usually spells its doom. A company that is without constancy of purpose does not think beyond the next quarterly dividend and usually has no long-range plan for staying in business. In other words, the company lacks a master plan. By adopting the new philosophy, management institutes leadership that people can follow. The PDCA cycle is at the heart of constantly improving the system.

Dedication to the new philosophy must be widespread throughout the company, and it is not sufficient to announce intentions to improve quality, however repeatedly. People have heard it before, many times over. They have seen many programs come and go in rapid succession—the so-called "program of the month." Dr. Deming went so far as to call this inconsistency of purpose the *crippling disease.* He said that we must start with a philosophy, a mission, and a vision for the future, around which we must build the master strategy. One way to demonstrate commitment is with money. Such concrete activities as spending money on training and equipment, or shutting down operations when something goes wrong, can be convincing, as can management taking the time to explain the new philosophy in full.

It is better to protect investment by working continually toward improvement of processes and of product and services that will bring the customer back again.

W. Edwards Deming
Out of the Crisis

Emphasis on Short-Term Profits

As Mary Walton so eloquently points out in *The Deming Management Method,* today's corporations are controlled by financial wizards and lawyers who manipulate numbers but do not make substantial changes in production and quality. They are subservient to stockholders, who are dependent upon them for the production, not of products and services, but ever-increasing quarterly dividends. Thus, it is common practice for companies to ship products on the last day of the month or quarter, without proper regard for quality. In fact, Dr. Deming says, in many cases they ship defective products in order to meet a quota. American management's emphasis on short-term profit is fueled by fear of an unfriendly takeover or leveraged buyout. "Where is the Securities and Exchange Commission?" says Dr. Deming. He calls this profiteering *paper profits* and points out that it has diverted attention and resources away from the job at hand—transforming the production base.

Must American management be forever subject to such plunder? Paper profits do not make the pie bigger. They give you a bigger piece but you take it from someone else. It doesn't help society.

W. Edwards Deming
The Deming Management Method

Evaluation of Performance Using Annual Reviews, Ratings, and Management by Objectives

Dr. Deming often referred to annual performance reviews, management by objectives (MBO), and management by numbers as a three-headed monster, which he called management by fear. He said that their effects are devastating in two respects. First, they encourage short-term performance at the expense of long-term planning and strategy. They discourage risk-taking and creativity, undermine synergy and teamwork, and pit people against one another for the same prize or rewards. People begin working for themselves, not the company, which results in competing or sparring fiefdoms.

Second, he said, "They focus on the end product, at the end of the stream, and not on leadership to help people. This is a way to avoid the people problems, and a manager becomes, in effect, Manager of Defects." A side effect, he points out, is that they increase reliance on numbers. Because they measure short-term results, the tendency is to rely only on evidence that can be counted, with little or no regard for the quality being produced. Merit ratings reward people who do well within the system and do not reward attempts to improve the system. "'Don't rock the boat' becomes the slogan," Dr. Deming said, thus cutting off creativity and risk-taking. Another problem, as he pointed out, is that they are useless as predictors of performance, except for those who fall outside the limits.

The final problem with MBO and performance appraisal schemes lies in the implied precision of rating schemes and systems. Dr. Deming felt that people or departments rated below average take a look at those rated above average and wonder why the difference exists. Thus, they try to emulate the people above average, which further impairs performance. This is further complicated by the inability of the supervisor to accurately assess workers fairly and equitably.

Management by the numbers leaves people bitter, crushed, bruised, battered, desolate, despondent, dejected, feeling inferior, some even depressed, unfit for work for weeks after receipt of rating, unable to comprehend why they are inferior. It is unfair as it ascribes to the people in a group differences that may be caused totally by the system that they work in....One gets a good rating for fighting a fire. The result is visible and can be quantified. If you do it right the first time, you are invisible. You satisfied the requirements, for that is your job. Mess it up and correct it later and you are the hero.

W. Edwards Deming
Out of the Crisis

Mobility of Top Management

Dr. Deming said that it is difficult for management to be committed to quality and productivity when the average tenure is only a few years. He pointed out the words of the managing director of JUSE who remarked: "America cannot make it because of the mobility of American management." Dr. Deming taught that the job of management is inseparable from the welfare of the company. "Mobility from one company to another creates prima donnas for quick results," he said, "for mobility annihilates teamwork, so vital for continued existence." Whenever a new manager comes in, everyone wonders what will happen next.

The mobility of labor in general is a problem for America in general and almost equal to the problem of the mobility of management. When people are dissatisfied with their jobs, they find it hard to take pride in their work. They are absent from work or start looking around for greener pastures. The problem starts and ends with management, whose job it is to encourage people to come to work and help them be productive and happy.

Unrest becomes rampant when the board of directors goes outside the company to bring in someone for a rescue operation. Everyone takes to the life preserver!

W. Edwards Deming
Out of the Crisis

Counting the Money

Dr. Deming believed that running a company on visible figures alone leads to failure. However important visible figures may be, the unknowns and intangibles must be taken into account. For example, consider the multiplying effect that a happy customer has on sales, as well as the opposite effect that an unhappy customer brings to the table. A happy customer who comes back for more is worth ten qualified prospects. "He comes without advertising or persuasion, and he may even bring a friend." He pointed to the story of the unhappy car buyer who, according to *Car and Driver* magazine (August 1993, p. 33), tells his troubles to an average of sixteen people.

In his discussions with Dr. Lloyd Nelson of Nashua Corporation, Dr. Deming mentions other examples of the "unknowables and unmeasurables" that lead to improvement of quality and productivity: improved upstream control, unshakable policy, better training and supervision, improved buyer and supplier relationships, and improved internal teamwork. He goes on to say that the costs of warranty are plainly visible, but do not tell the whole story about quality, because anybody can reduce these costs by refusing or delaying action on customer complaints. The real key is to avoid the reason for the complaint in the first place.

In *Out of the Crisis,* he tells a story about the credit department of a company that had put an emphasis on retaining only those customers who paid promptly, to the point that they often drove the others away. They performed well on the particular job assigned to them and by the numbers deserved a good rating. However, other less visible figures showed that they had driven some of the company's best customers to the competition by their intolerance of late payers. By the time top management looked at the total cost, it was too late.

He that expects to quantify in dollars the gains that will accrue to a company year by year, for a program of improvement of quality by principles expounded in this book, will suffer delusion. He should know before he starts that he will be able to quantify only a trivial part of the gain.

W. Edwards Deming
Out of the Crisis

Excessive Medical Costs and Obsolescence in Schools

Dr. Deming felt that excessive medical costs and a greedy system were choking the life out of the American economy. This may be a familiar refrain in Washington today, but Dr. Deming was saying it fifteen years ago, long before it was a popular political theme. In fact, he would say that medical costs remain the single biggest expenditure for some companies, comprising some thirty to forty percent of the annual operating budgets. Starting in medical school, students are taught that there is the "profession of business" and that they are ready to step into top jobs. Dr. Deming called this a cruel hoax. Most students have no experience in administration, sales, or the process of running a hospital or an office. Working in an emergency room at half the pay of a doctor just to get experience is a horrible thought to a professional administrator.

As time went on, Dr. Deming became more and more concerned about the crisis in American schools. When he last visited Miami in February 1992, we discussed the challenge outlined by the Commission on Educational Excellence for the Future. I asked him what he thought about the commission's warning, and he said that it had gone practically unheeded, even though it warned America:

> Our nation is at risk. Our once unchallenged preeminence in commerce, industry, science and technological innovation is being overtaken by competitors throughout the world. While we can have justifiable pride in what our schools and colleges have historically accomplished...the educational foundations of our society are presently being eroded by a rising tide of mediocrity that threatens our very future as a nation and as a people. If an unfriendly foreign power had attempted to impose on America the mediocre educational performance that exists today, we might well have viewed it as an act of war. As it stands, we have allowed this to happen to ourselves. We live among determined, well-educated, and strongly motivated competitors. We compete with them for international standing and markets, not only with products but also with the ideas of our laboratories.

Dr. Deming seemed particularly struck with the extraordinarily powerful final section written directly to students:

> You forfeit your chance for life at its fullest when you withhold your best efforts in learning. When you give only the minimum to learning, you receive only the minimum in return. Even with your parents' best example and your teachers' best efforts, in the end, it is your work that determines how much and how well you learn. Take hold of your life, apply your gifts and talents, work with dedication and self discipline. Have high expectations for yourself and convert every challenge into an opportunity.

Although the words were not his, the essence of the challenge is the same that Dr. Deming's teachings and philosophies deliver to students and businesses alike.

The best way for a student to learn a skill is to go to work in some good company (or hospital), under masters, and get paid while he learns.

W. Edwards Deming
Out of the Crisis

Excessive Costs of Warranty, Fueled by Lawyers Who Work on Contingency Fees

Dr. Deming said that America is one of the most litigious nations in the history of the world. In the business world, this is fueled on two fronts. First, he would argue, are the lawyers, whose gouging practices, as well as contingency and retainer fees, have substantially added to warranty costs of products and services. As if this were not enough, there are 10,000 accidental deaths at the workplace each year, for which the cost and loss are not countable. Add to this another 45,000 accidental deaths each year involving automobiles and the final picture is not a pretty one—almost 100,000 accidental deaths in America each year. Almost all of them come to involve lawyers, which adds untold millions to the cost of doing business and everyday living.

One of the documents we found interesting was *The Statistical Abstract of the United States,* which is available from the U.S. Government Printing Office. Among other things, it contains one statistic of which Dr. Deming took particular notice: the ratio of human resource outlays to national defense outlays, as measured by the annual federal budget. In 1945, the ratio was about one to fifty, or $1.859 billion for human resources to $82.965 billion for national defense, with a total outlay of $92.712 billion for these and other items. In 1950, when Dr. Deming first began his lectures, the ratio was about one to one; the total outlay was cut in half, but human resources had swollen to $14.221 billion and national defense was slashed to $13.724 billion. The human side of the equation had grown almost tenfold, a trend that would continue to its current levels. In 1993, human resources was $776.9 billion versus a national defense budget of $307.304 billion, or about a three to one ratio. In other words, using 1945 as the base year, human resource outlays rose over 100 times faster than national defense grew.

The implications, of course, are obvious. Two of Dr. Deming's deadly diseases—rising medical costs (#6) coupled with increased litigation (#7) surrounding human resource practices—have contributed significantly to the growth and escalation at an out-of-control pace. As we tracked these macro statistics on a year-by-year basis, it became obvious what needed to be done. The problem was, in the eyes of Dr. Deming, a leadership issue, starting with national politics and continuing all the way down to the local community.

Figures on accidents do nothing to reduce the frequency of accidents. The first step in reduction of frequency of accidents is to determine whether the cause of the accident belongs to the system or to some specific person or set of conditions. Statistical methods provide the only method of analysis to serve as a guide to the understanding of accidents and to their reduction...Accidents that arise from common causes will continue to happen with their expected frequency and variations until the system is corrected. The split is possibly 99 percent from the system and one percent from carelessness. I have no figures on the split, and there will not be any figures until people understand accidents with the aid of statistical thinking.

W. Edwards Deming
Out of the Crisis

Summary

In his never-ending effort to stamp out the Seven Deadly Diseases, Dr. Deming often was not gentle and would occasionally become impatient. His messages were warnings as he spotted the signs of infection and decay that only disease can bring. Above all, he wanted to be sure that his teachings were not compromised. None of the managers who ever attended his sessions misunderstood his message—for it was loud and clear.

When he was biting and critical, it was by design. He devoted his life to making this a better world, and it has been said that he used whatever means he could to communicate the new philosophy. He had seen so much so often that it is easy to understand why he became impatient and got upset. Sometimes he would get angry because American management had a hard time believing that there is a better way. Thomas Alva Edison, who many believe was the greatest mind ever to come out of America, had a saying above his office door: "There is a better way, find it!" Well, Dr. Deming, you found the better way and we are grateful to you for sharing it with us over the past 93 years of your life. And we are grateful to your family for sharing your time and presence with us. For none of us will ever be the same again.

You taught us well. And now it is up to us to carry on. May each of us be able to say when the end is near and our time has come, "I have done my best!"